U.S. Fire Administration

Funding Alternatives for Emergency Medical and Fire Services

FA-331/April 2012

U.S. Fire Administration

Mission Statement

We provide National leadership to foster a solid foundation for our fire and emergency services stakeholders in prevention, preparedness, and response.

Preface

The U.S. Fire Administration (USFA) would like to acknowledge the U.S. Department of Homeland Security's (DHS's) Office of Health Affairs for providing the substantial support necessary to perform this research and to develop this report.

This report was prepared through a cooperative agreement between the USFA and the International Fire Service Training Association (IFSTA) at Oklahoma State University (OSU). IFSTA and its partner OSU Fire Protection Publications has been a major publisher of fire service training materials since 1934 and, through its association with the OSU College of Engineering, Architecture, and Technology, it also conducts a variety of funded technical research on fire and emergency services and fire and life safety issues.

The extensive information provided in this report would not have been possible without the dedication and efforts of the following people assigned to this project:

- Project Administrator—Nancy J. Trench, Assistant Director for Research, Fire Protection Publications, OSU;
- Principal Investigator/Project Manager—Michael A. Wieder, Executive Director, IFSTA;
- Editor and Proofreader—Cindy Finkle, Research Assistant, Fire Protection Publications, OSU;
- Principle Writer—Scott Somers, FACETS LLP; and
- Principal Researcher—Kevin M. Roche, FACETS LLP.

The content of this report was reviewed by a group of subject matter experts (SMEs) and organizational representatives who have extensive knowledge and interest in this topic. The USFA and DHS's Office of Health Affairs would like to thank the following individuals and organizations for providing this oversight:

- Bill Troup—USFA;
- Richard Patrick—DHS, Office of Health Affairs;
- Dr. Jeffrey Lindsey, Ph.D.—Assistant Fire Chief, Alexandria Fire Department, VA;
- Bruce Evans—Emergency Medical Services (EMS) Chief, North Las Vegas Fire Department, International Association of Fire Chiefs (IAFC) Liaison to the National Association of EMS Physicians (NAEMSP);
- Mark Light—Executive Director/Chief Executive Officer (CEO), IAFC;
- Sean Caffrey—Interim EMS Operations Program Manager, EMTS Section, Colorado Department of Public Health and Environment;
- Philip Oakes—National Program Director/Trainer, National Association of State Fire Marshals (NASFM);
- Dennis Mitterer—MS, BSN, Emergency Medical Technician-Paramedic (EMT-P);
- Ronny J. Coleman—Representative for the National Volunteer Fire Council (NVFC);
- Dr. Lori Moore-Merrill—Assistant to the General President, International Association of Fire Fighters (IAFF); and
- Catherine Patterson—Branch Chief, Assistance to Firefighters Grants, Federal Emergency Management Agency (FEMA).

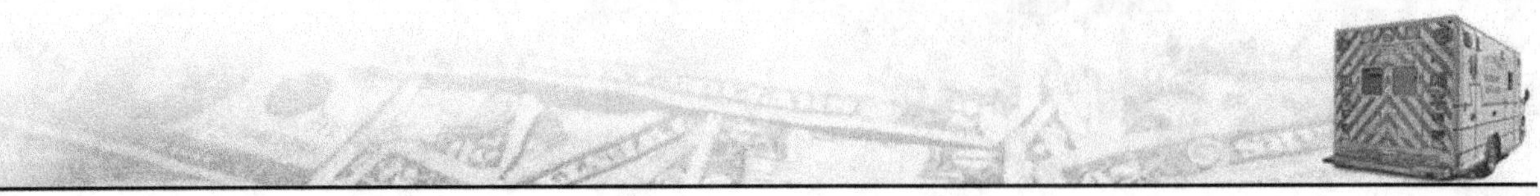

Table of Contents

Chapter 1: Introduction

Emergency medical services (EMS) departments and fire departments provide invaluable services in their communities. In the United States, EMS agencies and fire departments respond to millions of calls for service each year.[1] Service expectations placed on EMS and fire services organizations, including the fire service's role in EMS delivery, response to natural disasters, hazardous materials incidents, technical rescue, and acts of terrorism, have steadily increased.

This expanding mission comes at a time when local government budgets are stretched increasingly thin, which translates into more duties with fewer dollars to support them. A needs-assessment survey, conducted by the U.S. Fire Administration (USFA) in cooperation with the National Fire Protection Association (NFPA), documents a concerning lack of appropriate training, as well as inadequate facilities and equipment necessary to maintain basic fire and EMS response capabilities among the U.S. fire service, particularly in rural and volunteer fire departments. Few departments have the capability to handle unusually challenging events, such as wildland urban interface (WUI) fires, hazmat, or major floods.

Striking the right balance between different sources of local revenue has become an essential skill for EMS and fire administrators and chief officers. So, too, has the ability to identify and acquire grants and other alternative revenues.

To meet this critical need and serve its constituents regarding effective methods for obtaining funding, USFA developed the publication, *Funding Alternatives for Fire and Emergency Services* in 1993 and completed an update and revision of this document in 1999. This has been one of the USFA's most popular publications with thousands of copies being distributed and downloaded from the USFA website. In 2009, supported by the Department of Homeland Security (DHS) Office of Health Affairs, the USFA partnered with the International Fire Services Training Association (IFSTA) to update and revise the December 1999 edition of the report in order to provide the most up-to-date information regarding sources of funding for local-level EMS agencies and fire departments.

A key part of the project initiative was an enhanced study of critical funding issues for both fire and non-fire-service-based EMS systems. EMS is an essential public service in local communities. When illness or injuries strike, citizens depend upon EMS to respond with timely, high-quality care. According to the Institute of Medicine (IOM), more than 16 million patients arrive at the emergency room by ambulance each year. Even more are treated onscene without transport. A significant challenge for EMS is to ensure stable and sufficient revenue to provide for adequate prehospital care of the sick and injured. An optimal level of service in EMS typically presupposes revenue from more sources than taxpayer funding alone.

EMS agencies and fire departments require funding for expenses such as equipment, training, and salaries in order to provide necessary protection to their respective communities. However, with tighter budgets, less government subsidies, and fewer donations, it is becoming increasingly more difficult for EMS agencies and fire departments to meet greater and more complex demands for its services.

However, there have been many recent changes in the availability of Federal funding for EMS agencies and fire departments, including the Firefighter Investment and Response Enhancement (FIRE) Act, the Staffing for Adequate Fire and Emergency Response (SAFER) Act grant program(s), as well as other important grant programs. This updated document includes sources of Federal funding as well as other new and innovative funding sources not discussed in previous editions.

1 "Fire in the United States 2003–2007," 15th ed. USFA. October 2009.

Chapter 2: About This Manual

The purpose of this manual is to identify the various grants and innovative funding alternatives available for emergency medical services (EMS) and fire protection services. Where appropriate, a discussion of the pros and cons of a particular funding source is discussed. In some instances, examples are given to illustrate how alternative funding is being successfully implemented in EMS agencies and fire departments across the country.

This report is divided into the following several major sections identifying various sources of available funding and revenue options.

Local Revenue and Funding Alternatives

The funding alternatives available at the local level vary widely, though the ability of the EMS administrator or fire chief to influence changes in these sources will vary considerably based on local preferences and State and local laws. Many enterprising departments have found new revenue streams through the sale of services. The discussion of local revenue sources includes

- **Taxes:** These include taxes on real property, personal income, and sales transactions. Other taxes include real estate-transfer taxes and utility-user taxes. New taxes can go into a city's general fund to be divided up by elected officials or can be earmarked specifically for EMS and fire services.

- **Development Impact and User Fees:** These fees are charged to ensure that those benefitting from an activity pay their fair share of the costs related to that activity.

- **Fines, Forfeitures, and Citations:** Some jurisdictions now issue citations to those who engage in high-risk activities that may later require a rescue. Other areas dedicate a portion of fines to fund EMS and fire services.

- **Enterprise Funds and Utility Rates:** Local governments may establish an enterprise fund for city-operated services. Ambulance service, for example, may be run like a municipal business where it is expected to earn revenue to support its operations.

- **Sale of Assets and Services:** Some EMS and fire agencies sell used equipment or services to produce revenue.

- **Benefit Assessments:** Special districts may be established for the purpose of supporting EMS and fire services. These districts can assess a benefit assessment similar to a property tax but based on the "benefit" received by each property. These charges are a way to circumvent property-tax limitations and can also improve the equity of charges for EMS.

- **Borrowing:** Cities and towns have a number of options for borrowing revenue needed to purchase capital equipment and facilities including General Obligation bonds and Certificates of Participation. Not-for-profit organizations may have access to low-cost 501(c)(3) revenue bond financing or may take out a traditional bank loan.

This section also discusses creative ways fire chiefs and EMS administrators have raised revenue for their agency such as creating private ambulance company contracts to reimburse for fire-based EMS, billing for department-operated ambulance services, offering subscription programs, providing interfacility transport, creating paramedic intercept agreements with surrounding communities, and piloting innovating health-care programs. The section concludes with time-tested cost-saving strategies and miscellaneous fundraising ideas for rural and volunteer fire and EMS agencies.

State and Federal Funding Programs

In addition to sources of local revenue derived from taxes and other local government financial mechanisms, EMS agencies and fire departments may be able to obtain considerable funding from State and Federal programs. Two chapters list grants available from States and the Federal government.

Often, Federal grant funding flows to the local level through the States. However, some States provide direct funding for EMS, especially in rural areas. Many States offer grant aid and low-interest loans for capital-improvement projects. On the other hand, some States have no funding for local programs. Most State Fire Marshal's Offices and EMS bureaus offer technical assistance to local agencies and subsidized training programs to first responders.

EMS administrators and fire chiefs in States without funding may wish to consider what other States are doing and determine if some of those methods might be feasible in their State. The taxpayers pay for EMS one way or the other, but some approaches may be more palatable or more equitable than others.

Agencies of the Federal government offer thousands of grants and special programs to State and local governments; many of them may apply to EMS. Some programs are not specifically earmarked for fire or EMS but can be used for those purposes, such as community block grants.

Private-Sector Sources

Private foundations and corporate-giving programs donate millions of dollars annually for education, arts, and community development, among other projects and programs. Career and volunteer EMS and fire agencies may be able to raise funds for capital purchases or to improve service delivery through these sources, especially in rural communities or poor urban areas. When looking at these options, caution should be taken to ensure that accepting donations from these types of programs does not result in the appearance of a corporate preference or conflict of interest. This is of particular importance for EMS and fire organizations that are government-funded. This is of lesser concern for volunteer organizations that operate outside the direct control or funding of their local government. Some public EMS and fire departments establish separate 501(c)(3) nonprofit organizations to accept donations for local causes related to their mission. Among the private sector sources of giving are

- **Foundations:** Some communities are fortunate to have community/public foundations whose funds can be used for providing special public safety services, starting new programs, helping low-income areas, or other services allowed by charter. There are also national independent and corporate foundations that donate to communities across the country. Foundations provide a source of revenue through general operating support grants, planning grants, seed money, management assistance grants, and facilities and equipment grants.

- **Corporate-Giving:** Many public safety agencies have been successful in soliciting grants and services from national corporations. Companies are often interested in providing community assistance in cities and towns where they have a corporate presence or where their employees live. Corporate-giving has public relations benefits and emergency service administrators should be ready and willing to recognize corporate sponsors whenever possible. Corporate-giving comes in the form of cash donations, in-kind contributions, and even executive loan programs.

- **Program-Related Investments:** In addition to grants, cash, or in-kind donations, foundations and corporations may be a source of low-interest or zero-interest loans, particularly for capital-improvement projects.

Writing an Effective Grant Proposal

The final chapter of this manual presents fundamentals of developing grant proposals to assist EMS administrators and fire officers in preparing appropriate documentation in pursuit of grants from government and private sources. While this advice is generally applicable to any grant application, grant seekers must always read, understand, and follow the specific program guidelines for grant opportunity.

Implications of Funding Choices

Each agency providing EMS and fire protection must consider the range of opportunities and the local and State constraints in shaping its funding policy and determining its budget. Funding influences the amount of response resources available and the scope of preventative activities which, in turn, influences patient outcomes. The funding issue is not one to be left solely to accountants and financial officers; it is a vital public policy issue that deserves community input.

The choice of funding approaches also raises fundamental questions about governance and equity. Many questions about EMS have become part of the larger national health-care debate: Is EMS solely a private good where users of prehospital medical services pay for the burden of the services provided to them? Or, is EMS a public good provided by the community to people in need, and the cost spread across society? Perhaps EMS should be treated like insurance, in which the fees charged are a function of the risk presented? Who should provide EMS?

These are fundamental questions about our values and the role of government in society and apply to the fire service as much as they do EMS or any service the government provides. Different communities are taking radically different approaches; many seem uneasy about charging for services beyond the taxes paid, while others have found it to be the only way to fund critical public safety services.

Providers of emergency services should consider the wide variety of funding alternatives available to finance services and perhaps improve the equity of paying for the service. Citizens often balk at new fundraising mechanisms. Changing the way public safety is funded can profoundly affect the way government is viewed, and change the unwritten contract between a government and its citizens on major issues such as representation and the purposes and goals of government. EMS and fire leaders must take a collaborative and participatory approach to addressing complex public policy issues, including discussions about the role of EMS and fire agencies in the community and how public safety services are funded.

Finally, local government charters and State constitutions may limit the extent to which EMS agencies and fire departments can generate revenue. Some States prohibit volunteer ambulance agencies from billing for services. Others have restricted the amount of development impact fees that can be charged and prohibited the adoption of new user fees. Before developing any new cost-recovery program, EMS managers should consult with an attorney to discuss any limitations.

Scope

This manual discusses alternative sources of revenue available to fund EMS and fire protection in all types of systems: career or volunteer, urban or rural, advanced life support (ALS), or basic life support (BLS). While fire-based EMS agencies and private-sector ambulance companies are an important part of many systems, some of the grants and funding sources contained herein apply only to the fire service. The fire service is the primary first responder in many communities and many fire departments play a major role in providing EMS. Currently, fire departments with cross-trained dual-role personnel constitute 38 percent of EMS responders in the United States, more than any other single-system type.[2] Many of the Federal grants autho-

2 "EMS By the Numbers," National Association of Emergency Medical Technicians (NAEMT).

rized by Congress since 9/11 focus on enhancing the capabilities of first responders to respond to terrorist events and incidents of national significance, particularly in law enforcement and fire departments. Regardless, non-fire-based EMS agencies should find many of these innovative funding alternatives applicable to their organization.

Sources of Information

The birth of the Internet era has produced an "information superhighway" providing users with timely access to information. Nearly every grant-making agency, public or private, has a web presence; though some are more conspicuous than others. Websites contain far more information about a grant than can be included in this manual. It is recommended that researchers use this manual to locate grants and gain ideas about what funding alternatives might be available, then conduct additional research on the web or at the local library.

Several websites were particularly useful in compiling this manual. Grants.gov (www.grants.gov) and the Federal Grants Wire (www.federalgrantswire.com) are free resources that index thousands of grants and loan programs from nearly every Federal agency. The Federal Emergency Management Agency (FEMA) maintains a website for the Assistance to Firefighters Grant program (www.fema.gov/firegrants) with details about each grant, grant writing tips, workshop schedule, and online application. The Foundation Center (www.foundationcenter.org) profiles private-sector resources as well as provides online tutorials on grant-writing skills. Finally, State agency websites contain information about passthrough grants managed by the State or State-funded programs that might be useful to EMS providers.

Uniform Resources Locators (URLs) to funding source webpages are included whenever available. It should be noted, however, that while these URLs are provided for the convenience of the reader, webpages are subject to link-rot over time. That is, links pointing to webpages, servers, and other Internet resources may become permanently unavailable or out of date. If this happens, it is recommended that the researcher attempt to use an Internet search engine to search the name of the grant or contact the grant-making agency directly to inquire if the grant is still available.

In addition to a review of websites and databases, a detailed bibliographic search was undertaken. Project staff, staff of the U.S. Fire Administration (USFA) and International Fire Service Training Association (IFSTA), and the peer-review team also contributed many ideas. Some success stories were obtained from contacting past grant recipients to provide a brief synopsis of their project.

Every effort was made to compile a comprehensive listing of grants and programs available to EMS agencies. In doing so, however, some programs may have been overlooked. They either were not known to those persons contacted or otherwise not included in the literature or websites reviewed for this project. It is also possible that grants and programs listed here are no longer available. Grant seekers should contact grant-making agencies directly to find out which programs are currently funded and what funding may become available in the future. Regardless, this manual should provide a good starting point for EMS managers seeking information on innovative solutions and alternative funding sources to support EMS in their community.

Summary

Historically, fire protection and EMS provided by fire departments have been funded as part of the municipal budget; only ambulance companies billed for their services. Pressured by the tax revolts of the 1970s and 80s and the fiscal crises of the early 1990s and late 2000s, EMS and fire administrators are thinking like entrepreneurs, increasing organizational efficiency, and seeking out new revenue streams. Changes in health-care financing and new legislation are challenging, traditional approaches to managing EMS. These

pressures are particularly difficult for EMS and fire providers as they come at a time when there are increased demands for the services they provide and a need to add capacity in order to prepare for emerging threats to public health and safety, such as terrorism and pandemic.

Public safety providers need to know the range of possible funding alternatives available to them, especially those that have proven effective. This manual describes alternative sources of funding and revenue-producing opportunities that may be used to finance EMS and fire protection systems.

Chapter 3: Writing an Effective Grant Proposal

Successful grant writing involves thoughtful planning and preparation. It takes time and persistence to succeed. A successful grant is one that is well-prepared, thoughtfully planned, and concisely packaged.

Whether pursuing a government grant or private dollars, it is important to read and understand the specific program guidelines for that grant opportunity. Always follow the exact specifications of the grant in the application. Every grant or foundation provides information on program priorities and the information needed to include in an application in order to be successful. Failing to request top-priority items is the number one reason why applications are unsuccessful. Reading the guidelines will give the applicant a good idea of what information is needed to include in the application narrative.

Government agencies and large foundations may have formal application packets, strict guidelines, and fixed deadlines that must be adhered to. However, the steps outlined here can generally be applied to any grant-seeking effort. This is not a rigid formula; rather it is a suggested approach that can help the grant-seeker get organized and write a winning grant proposal.

Grant Fundamentals

For many, grant writing can be an intimidating prospect. But, with a little diligence, the rewards can help fund projects important to the safety, health, and welfare of the local community. So, where does one start? The following are five important grant fundamentals to consider when planning a grant application.

1. Identify Grant Opportunities First.

Too often, grant-seekers get caught up identifying and ranking agency needs, then searching out grants for their top priorities. Does it matter that getting a grant for an automatic external defibrillator (AED) may not be at the top of the list? If a grant is available for the AED, why not apply? Rather than looking for grants to fund a particular project, look at grants and see if they meet an agency need. Try to align the agency's needs with national priorities and where the money is located.

Searching for grant opportunities has never been easier. This guide contains hundreds of government grants and private foundation funding sources with important information such as a grant synopsis, contact information, and websites. The advent of the Internet has made searching for grants even easier. Research available grant opportunities and match funding to an organizational need.

2. Research Opportunities.

Once you have identified a grant and matched it to an agency need, it is important to fully understand the nature of the grant. Failing to request top-priority items in a grant is the number one reason why applications are unsuccessful. It does not matter how well written an AED grant application is, if it is submitted to a grant that focuses on paramedic training, the grant will almost assuredly be rejected.

Because most grants are competitive, it is important to understand the grantor's priorities and match the department's needs with those priorities. The key to successfully obtaining grants is to read the grant guidance. Pay particular attention to the funding priorities, specified application procedures and documents, signature requirements, and deadlines.

Before applying for a grant, ask the following questions:

- Is the agency eligible to apply?
- Is the project eligible for consideration in terms of goals, timing, and type of activity?

- What are the grantor's priorities?
- Who and what projects have they funded in the past?
- Which expenses can be funded by the grant?

3. *Write a Good Narrative.*

The information provided in the narrative is what the peer reviewers use to determine if they recommend your project for funding. A poorly written or confusing narrative is a quick path to rejection. The grantor must be able to read and understand what it is the department is applying for and what it's trying to accomplish. At a minimum, the narrative should provide a project description, including community risk and budget, an explanation of the financial need, a description of the cost-benefit of the proposed project, and an explanation of how the project will improve daily operations.

At some point, perhaps after the first or second draft is completed, seek out a neutral third party to review the proposal-working draft for continuity, clarity, and reasoning. Ask for constructive criticism at this point, rather than waiting for the grantor agency to volunteer this information during the review cycle. For example, has the writer made unsupported assumptions or used jargon or excessive language in the proposal?

4. *Follow All Instructions Carefully.*

Before submitting the final package, verify that all of the instructions have been followed carefully, all of the requested documentation has been included in the application packet, the information is accurate, and required signatures are included. Some Firefighter Investment and Response Enhancement (FIRE) Act grant applications are rejected because they lack the required documentation or signatures. Be sure to keep a copy of the submission for the agency's records. Above all, do not miss the deadline; it is the only way to know for sure the application will be considered.

Many grantor agencies now offer online application submittal. Use this process whenever possible. Online applications dramatically simplify the process and they do not allow the writer to submit an incomplete application. Be sure to print a hard copy of the application and save the email confirmation receipt.

Following all instructions also includes complying with reporting requirements. Nearly all grants require some form of progress report and closing report. Grantors are primarily interested in how the funds were used to meet the department's needs and ensure they match the requirements of the grant. Remember, misappropriation of grant monies can be considered theft and subject to civil and criminal prosecution.

5. *Get Help.*

Grant writing can be a daunting task. Don't be afraid to ask for help. There are many ways to become educated on grant writing and grant management. Online tutorials are available and many private agencies and academic institutions offer classes and workshops. The Foundation Center (www.foundationcenter.org) is a leading source of information on philanthropy, fundraising, and grant programs. It offers online resources and free assistance at any of its regional offices.

Basic Components of a Proposal

In general, the basic components of a standard grant proposal include the following:

- cover letter;
- executive summary or abstract;

- introduction describing the grant-seeker or organization;
- problem statement or needs assessment;
- goals and objectives;
- methods;
- evaluation;
- sustainability; and
- project budget.

Cover Letter

The cover letter should provide a clear, concise overview of the organization, project description, and financial need. It should be written on letterhead, signed by the organization's highest official, and addressed to the individual at the funding source with whom the organization has dealt. Be sure to refer to any previous discussions with the grant manager. The cover letter should demonstrate a familiarity with the mission and goals of the grant-making agency or foundation, and emphasize the ways in which the application contributes to these goals. Cover letters should be no more than one page.

Executive Summary (or *Abstract*)

Also referred to as a cover sheet, the purpose of the executive summary is to provide the reader with an overview of the longer grant application. Think of it as a condensed summary of the entire application. It is designed to persuade the readers that the project is a winner and should be funded.

The executive summary should include a description of the applicant and contact information, a definition of the problem to be solved, a statement of the objectives to be achieved, an outline of the activities and procedures to be used to accomplish these objectives, a description of the evaluation design, plans for the project at the end of the grant, and a statement of what it will cost the funding agency. It should also identify other sources of funding or collaborative partnerships involved in the project.

The executive summary should be brief—no more than two or three paragraphs. Resist the temptation to pad the executive summary with details. The purpose of the summary is to present the facts and entice the reviewer to read the entire proposal. Keep the language strong and positive. Tailor the information to the audience. Technical language is fine for a peer-review process, but may be too much in a philanthropic grant review.

Remember, the executive summary is the first thing the grant reviewers will read. If it is poorly written, it may be the last thing they read and the project will not be funded.

Introduction

In the introduction, applicants describe their organization and demonstrate that they are qualified to carry out the proposed project; they establish their credibility and make the point that they are a good investment.

Most proposals require a description of an applicant's organization. In two pages or less, provide a brief history of your organization, call volume, current budget and funding challenges, problem or risk, and proposed project solution. Highlight the organization's mission and goals and be sure to demonstrate how the subject of the grant proposal fits within or extends those mission and goals.

Discuss the organization's structure. If the organization has a board of directors, describe how board members are recruited and their level of participation. If the organization has an active volunteer group, describe the function volunteers perform. Provide qualifications of the professional staff related to the grant application and their levels of expertise.

Describe your community characteristics and briefly explain any special services targeted towards at-risk groups. Describe the constituency served, any special or unusual needs they face, and why they rely on agency programs. Cite the number of people who are reached through organizational programs.

Problem Statement (or Needs Assessment)
This section lays out the reason for the proposal. It should clearly and concisely demonstrate that a relevant, compelling problem or need exists. Whenever possible, include qualified third-party research/evidence to help justify the need.

The best way to collect information about the problem is to conduct a needs assessment. The information provided should be both factual and directly related to the problem addressed by the proposal. Areas to document include

- Purpose for developing the proposal.
- Beneficiaries—who are they and how will they benefit?
- Social and economic costs to be affected.
- Nature of the problem (provide as much hard evidence as possible).
- How the applicant organization came to realize the problem exists and what is currently being done about the problem.
- Stress the gaps that exist in addressing the problem that will be remedied by the proposal.
- Sustainability—explain how the program will continue to benefit the community after the grant funding has been exhausted.
- Most importantly, specify the manner through which problems might be solved. Review the resources needed, considering how they will be used and to what end.

One of the pitfalls to avoid is presenting the absence of the solution as the actual problem, or circular reasoning. For example, the lack of a Spanish-speaking drowning prevention program is not a problem—the problem is that the community has a high rate of child drowning in Spanish-speaking communities. A persuasive argument would document the child-drowning data in the community in comparison to English-speaking areas. Case studies or academic publications may be cited to demonstrate how similar programs have been successful in other communities in reducing child drowning.

Goals and Objectives
Setting goals and objectives is an important part of project planning and grant writing. Goals are statements about general aims or purposes of a program. An objective is a precise, measurable statement of what a program intends to achieve during a specific period of time toward a particular desired outcome.

Goals are abstract, whereas objectives are specific statements that state who will make what change, by how much, where, and when. A commonly used acronym for writing program objectives is SMART: specific, measurable, achievable, realistic, and time-specific.

Remember, if the proposal is funded, the stated objectives will probably be used to evaluate program progress, so be realistic.

Methods

The objectives explain to the grantor agency what will be achieved by the project. The methods section describes the specific activities that will take place to achieve those objectives. A specific plan (or method) should be laid out for each objective in the grant proposal. The methods section should describe how, when, and why.

- **How:** Describe the process to be used to achieve the objectives in a rational, direct description of the actions that will accomplish the objectives. Be sure to include innovative features of the proposal which could be considered distinct from other proposals under consideration. The methods must match the stated objectives.
- **When:** Present a timeframe for the grant project or program. When will the project begin? When will each step be completed? It may be helpful to reviewers to present a visual version of the project timeline to help them understand and evaluate the planning and feasibility.
- **Why:** Wherever possible, justify the course of action to be taken. Why will the work plan best meet the objectives? If requesting funding for a product, why will that product best work to accomplish the objectives? Including case studies and expert testimony as part of the grant application are ways to help answer these questions.

Projects can be easily laid out using off-the-shelf illustrator software programs that run on any personal computer. A Program Evaluation and Review Technique (PERT) chart could be useful in justifying some proposals. This software can be used to create a PERT chart that provides a graphical representation of all tasks in a project and the way tasks are related to each other.

The methods section should include

- A restatement of the problems and objectives.
- A clear description and explanation of the project scope and activities.
- Activities to occur including the related resources and staff needed for the project.
- A flowchart of the organizational features of the project describing how the parts interrelate, where personnel are needed, and what they are expected to do.
- Timeline of activities.
- Whenever possible, use appendices to provide details, supplementary data, references, and information requiring more detail.

Evaluation

Grantor agencies want to know that they are getting their money's worth and that grant money was spent properly. An evaluation plan should be built into the project. Including an evaluation plan in the grant application indicates that the grantee takes the objectives seriously and wants to know how well they have been achieved.

There are two types of formal evaluation: product and process. Product evaluation addresses the result that can be attributed to the project, as well as the extent to which the project has satisfied the stated objectives.

Process evaluation addresses how the project was conducted, in terms of consistency with the stated plan of action and the effectiveness of the various activities within the plan. Either or both might be appropriate to a project. The evaluation approach will depend on the nature of the project and its objectives.

Most grants contain some form of program evaluation or reporting requirement. The requirements of the proposed project should be fully understood before applying. Evaluations may be conducted by an internal staff member, an external auditor, or both. The applicant should state the amount of time needed to evaluate, how the feedback will be distributed among the proposed staff, and a schedule for review and comment for this type of communication.

Evaluation requires both coordination and agreement among the program decisionmakers. Above all, a grantor agency's requirements should be highlighted in the evaluation design. Some agencies may require specific evaluation techniques such as designated data formats (such as an existing information-collection system) or they may offer financial inducements for voluntary participation in a national evaluation study. Typically, Federal grants will have a description of the exact evaluation methods and reporting requirements in the grant's program guidance documents.

Sustainability

A clear message from grantor agencies today is that grant-seekers will be expected to demonstrate how the overall activity can be sustained beyond the grant performance period and that the project has potential long-term benefits. The narrative should describe a plan for continuation beyond the grant period and outline all other contemplated fundraising efforts and future plans for applying for additional grants. Projections for operating and maintaining facilities and equipment should also be given.

Project Budget

The budget for a proposal can be as simple as a one-page statement of the projected revenue and expenses. A proposal may, however, require a more complex presentation, perhaps including a page on projected support and notes explaining various items of expenses. If costs are straightforward, a simple spreadsheet may suffice. Otherwise, use of a narrative format to explain unusual line items in the budget may be necessary. The narrative should be succinct, dealing only with the budget, not a restatement of the project. Alternatively, consider using footnotes attached to specific line items needing more explanation.

Generally, budgets are divided into two categories: personnel costs and nonpersonnel costs. In preparing the budget, the applicant may first review the proposal and make lists of items needed for the project. The personnel costs include

- salaries (including increases in multiyear projects);
- benefits, such as health insurance and retirement plans; and
- consultants and contract services.

The items in the nonpersonnel costs include

- space/office rental and leasing costs;
- utilities;
- equipment purchase or rental;
- training to use new equipment;

- travel expenses; and
- office supplies (photocopying, computer supplies, etc.).

The budget proposal should also account for in-kind contributions. These services or materials can oftentimes be used as a "match" for many grants. Examples of in-kind contributions include

- volunteers and probono services;
- use of building space and utilities;
- transportation;
- printing and advertising costs; and
- insurance.

A well-prepared budget justifies all expenses and is consistent with the proposal narrative. Some areas in need of an evaluation for consistency are

- Salaries of the proposal in relation to those of the applicant organization should be similar.
- If new staff persons are being hired, additional space and equipment should be considered, as necessary.
- If the budget calls for equipment purchase, it should be the type allowed by the grantor agency.
- If additional space is rented, the increase in insurance should be supported.
- In the case of Federal grants, if an indirect cost-rate applies to the proposal, the division between direct and indirect costs should not be in conflict, and the aggregate budget totals should refer directly to the approved formula.
- If matching funds are required, the contributions to the matching fund should be taken out of the budget unless otherwise specified in the application instructions.

If pursuing a Federal grant, it is very important to become familiar with government guidelines for Federal domestic assistance. In learning to develop a convincing budget and determining appropriate format, reviewing other grant proposals is often helpful. The applicant may ask the grantor agency for copies of winning proposals. Appendix A provides an example of a successful grant application in support or rural emergency services.

Where to Get Help

Grant writing can be confusing and takes time and expertise. Applicants should consider taking a grant workshop to strengthen grant-writing and grant-management skills. The Federal Emergency Management Agency (FEMA) webpage for Assistance to Firefighter Grants (AFG) (www.fema.gov/firegrants/) offers online grant-management tutorials. The tutorial provides step-by-step instructions on how to manage grants. FEMA also sponsors grant-management workshops for applicants to learn the details about specific grants and talk directly to AFG regional contacts.

Colleges and universities typically offer grant-writing workshops. These programs can help those who wish to strengthen their grant-writing skills to acquire and master the techniques of preparing, writing, and

winning proposals from various funding sources. These workshops focus on the basics of the grant-writing process, effective ways to write and prepare grant applications, finding sources of funding, and other relevant topics.

Many municipalities hire a grants coordinator to perform grant-seeking, grant administration, and grant-coordination activities across city departments. Grant coordinators may perform any number of the following tasks:

- Research and identify Federal, State, and private grant-funding opportunities.
- Develop and catalog external resources in the areas of training, grant-funding sources, and other grant-related resources.
- Provide assistance in determining grant eligibility.
- Provide technical assistance and training to departments in grant writing, preparation, and management.
- Develop and implement administrative procedures and controls for grant-application process.
- Facilitate interdepartmental coordination and communication on multidepartmental grant projects to ensure that grants are processed and administered in a timely manner and according to grant requirements.
- Provide technical expertise in determining grant-compliance requirements.

Smaller departments and volunteer organizations might consider hiring a grant consultant. Consultants typically charge an hourly fee. Some grant writers charge according to the size of the agency. For example, those with a budget of $250,000 or less get a 10- to 15-percent reduction in the hourly rate charged to agencies whose budgets are bigger. In any case, hiring a consultant may be worth the expense if it means the difference between receiving a grant or not.

When hiring a grant consultant, be sure the applicant has a strong understanding of the grant-life cycle process including grant-seeking, grant writing, and the grant-management process. Practical experience in these areas is highly desirable. The applicant should demonstrate considerable experience in research, finance, or public administration, including obtaining funding for grant projects and programs, and grant writing.

Interaction and Networking

Completing a grant proposal is a major step to obtaining a government grant or foundation funding. Interaction with these organizations may assist an emergency medical services (EMS) agency or fire department in their quest for grant funding, as well as help them identify when opportunities arise for grants and when the "time is right" for grant proposals. Interaction and networking are key elements in marketing the department, the grant proposal, and finding new funding sources.

State Agencies

Each State EMS bureau, fire marshal's office, or Homeland Security/Emergency Management agency is responsible for the direction of public health and safety at the State level with requirements filtering down to the local level. Most State authority is supported by State legislation which enables the agency to carry out its responsibilities, establishes limits of authority, and provides some appropriations from State funds. The

State office may be a separate authority or it may be housed within another State agency, such as the State health department or highway/transportation department. Functions of the State office include

- regulation of provision of ambulance or medical transportation and fire protection services;
- establishment of training standards for EMS and fire personnel;
- interoperable communication systems;
- disaster planning, response, and financial assistance;
- data collection; and
- system evaluation.

It is important to understand the political authority and activities of the State EMS, fire, and Homeland Security offices in order to successfully participate in the system at the local level. Because most State EMS offices are within the State's public health system, authority, and legislative initiatives should also take a public health or "preventive" approach. When attempting to receive funding from State offices, it is essential to understand the laws and associated regulations that have an impact on particular areas of the profession. To increase effectiveness when dealing with the State EMS, fire, and Homeland Security offices, agency leaders should know the answers to the following questions:

- What does the State EMS, fire, and/or Homeland Security office do?
- How are its activities authorized?
- Who is involved with implementation?
- What other statutes or regulations exist?

Other Associated Organizations

In addition to understanding the State's lead EMS, fire, and Homeland Security agencies, it is also important to be aware of other influential organizations. Lead agencies in every State exchange information and services with many other organizations, agencies, and support groups. One entity—and an integral part of the EMS system structure—is the State-level advisory board or council. Although considered to be advisory in nature, this board or council is usually politically strong, and its recommendations to the State EMS lead agency can significantly impact the direction of the EMS program. Again, effective preparation for influencing this agency involves researching the answers to important questions:

- When and where does the council meet?
- Who are the members?
- What EMS areas do they represent?
- How often are they appointed and by whom?

Other organizations that can affect change in the EMS arena are State affiliates of various EMS organizations, the State board of nursing, and the State medical society. Also, in addition to the State lead agency, there are several related State agencies to consider in EMS initiatives. Each State's governor's office has an Office of Highway Safety, which has a similar agency at the national level, the National Highway and Traffic Safety Administration (NHTSA) of the Department of Transportation (DOT). This national office provides limited

funding for EMS initiatives. Since the State Office of Highway Safety represents the governor's office, it becomes a valuable link in the network chain, not only for financial purposes but also for its political support. Spin-off programs administered by this office include passenger restraints, motorcycle safety, alcohol and drug use, and impaired driving programs, all of which are related to improving emergency medical care. New program initiatives, legislation, and other activities should include the involvement of these offices from the beginning.

Books on How to Write a Grant Proposal

If you need help writing a grant proposal, these selected handbooks can help applicants become successful grant writers. Check your local library for these and additional resources.

The Complete Book of Grant Writing: Learn to Write Grants Like a Professional. Nancy Burke Smith and E. Gabriel Works, 2006. This book covers every section of grant proposal, instructs how to make proposal compelling, explains timetable. Includes sample proposals, letters of inquiry, and support letters.

Complete Idiot's Guide to Grant Writing. Waddy Thompson, 2007. This overseer of fundraising activities for a foundation describes each step of the process including how to determine who can receive a grant and for what, how to find donors, how to create a budget, how to write a compelling proposal, and how to overcome obstacles.

Demystifying Grant Seeking: What You Really Need to Do to Get Grants. Larissa Golden Brown and Martin John Brown, 2001. Practical advice on building a grant-seeking process for your organization beginning with dispelling myths/fears that get in the way of effective grant-seeking and continuing with entire grant-seeking process improvement.

The Everything Grant Writing Book. Judy Tremore and Nancy Burke Smith, 2008. This book discusses all aspects of grant-writing process to help you succeed in competitive world of grant funding.

Foundation Center's Guide to Proposal Writing, 1997. A manual on basic, finer points of grant-seeking. Covers proposal planning, writing, and postgrant followup. It includes actual examples, excellent insider information, and tips from grant-givers in question-answer format.

Foundation Fundamentals: A Guide for Grant Seekers. Mitchell F. Nauffts, 1994. A comprehensive, easy-to-read guidebook with facts needed to understand world of foundations. It helps identify appropriate foundation funding sources and provides detailed instructions on researching private foundations.

Grant Application Writer's Handbook. Liane Reif-Lehrer, 2005. This was written by an author who's been on the receiving end of many research grants. It covers all steps and features helpful forms and tips.

Grant Proposal Makeover: Transform Your Request from No to Yes. Cheryl A. Clarke and Susan P. Fox, 2007. Covers how to edit a proposal to improve chances of success. Insider tips from funders.

Grant Writing for Dummies. Bev Browning, 2009. A step-by-step guide to writing winning proposals. This text covers grant types, terminology, explanation of failure/success of proposals, samples, and etiquette for following up.

How Foundations Work: What Grantseekers Need to Know about the Many Faces of Foundations. Denis McIlnay, 1998. An inside look at foundations. The author debunks myths and provides productive strategies for dealing with foundations.

The "How To" Grants Manual: Successful Grantseeking Techniques for Obtaining Public and Private Grants. David G. Bauer, 2003. This contains detailed, experienced information on every step of the grant-application process and best methods for approaching foundations, corporations, and government funding sources.

How to Develop and Write a Grant Proposal. Merete F. Gerli, CRS Report for Congress, 2007. Order Code RL32159. This report is intended for members and staff assisting grant-seekers in districts and States and covers writing proposals for both government and private foundations grants.

How to Write a Grant Proposal: Everything You Need to Create a Winning Proposal. Matthew Lesko and Sarah Priestman, 2004. This explains Lesko's 10-step program for writing a winning grant proposal.

How to Write a Grant Proposal. Cheryl Carter New and James Aaron Quick, 2003. Writing a proposal is the end of a process, not the beginning. This book concentrates on behind-the-scenes, prewriting work that makes the difference between winning and losing proposals. Guidance for educational institutions, government agencies, for-profit businesses, traditional nonprofits, and others.

How to Write and Get a Grant. Matthew Lesko and Mary Ann Martello, 2002. This outlines government and nonprofit grant sources and provides advice on preparing applications and proposals.

The Only Grant Writing Book You'll Ever Need. Ellen Karsh and Arlen Sue Fox, 2006. This book treats grant writing as the final, not-too-difficult step in a comprehensive, methodical process.

"Thank You for Submitting Your Proposal": A Foundation Director Reveals What Happens Next. Martin Teitel, 2007. The do's and don'ts drawn from 30 years of experience in the funding world. The text includes "Grant Seeker's Reality Check" with concise, easy-to-grasp advice.

Webster's New World Grant Writing Handbook. Sara Deming Watson, 2004. Covers every step of proposal development, as well as what should happen after it's sent.

Winning Grants Step by Step: Support Centers of America's Complete Workbook for Planning, Developing, and Writing Successful Proposals, 2002. An excellent workbook for planning and composing grant proposals. It includes exercises to develop skills and meet requirements of both government agencies and private funders.

Help for Nonprofit Organizations

Earning More Funds: Effective, Proven Fundraising Strategies for Every Non-Profit Group. Chip and Ralfie Blasuis, 1995. All types of nonprofit fundraising ideas plus short, useful overview of grant-writing fundamentals.

Everything Fundraising Book: Create a Strategy, Plan Events, Increase Visibility, and Raise the Money You Need. Rich Mintzner and Sam Friedman, 2003. This includes sections on "All About Grants" and "Grant Writing."

Everything Guide to Fundraising: From Grassroots Campaigns to Corporate Sponsorships—All You Need to Support Your Cause. Adina Genn, 2009. Contains details about organizing a fundraising campaign. "All About Grants" and "Grant Writing 101" sections are also included.

Fundraising for Non-Profit Groups: How to Get Money from Corporations, Foundations, and Government. Joyce Young, 2002. A discussion of funding and helpful descriptions of entities who supply money for nonprofit organization work.

Where to Find Help on the Internet

"Developing and Writing Grant Proposals"
www.vfda.gov or http://purl.access.gpo.gov/GPO/LPS1729

This is an instructional document on Catalog of Federal Domestic Assistance website. Look for "Writing Grant Proposals" in lower left menu.

Education World–The Grants Center
www.educationworld.com/a_admin/archives/grants.shtml

Grant information for educators including grant writing resources, guides. Available grants/grant originators are arranged by subject.

EMS Grants Help
www.emsgrantshelp.com

A resource for EMS, ambulance, and paramedic grants. It provides members of EMS with the most comprehensive resource for EMS grant listings, information, and resources. Registration is free.

Federal Grants Wire
www.federalgrantswire.com

Federal Grants Wire is a free resource for Federal grants, government grants, and loans. The site currently indexes over 2,400 Federal grants and loans organized by sponsoring agency, applicant type, subject area, and a convenient directory to begin a search. The site includes multiple resources which provide information such as how to write a grant proposal, tips on applying for Federal grants, application eligibility requirements, and more.

Fire Grants Help
www.firegrantshelp.com

Fire Grants Help provides firefighters and fire departments with a comprehensive resource for grant information and assistance. It features a grant database and grant-assistance tools such as research guidance, grant-writing tips, grant resources, and more.

The Foundation Center
http://foundationcenter.org

This site includes many features, some free and others fee-based. Free features include "Foundation Finder" with profiles of foundations (searchable by name or part of name, State, or zip) and "990 Finder" allowing review of Internal Revenue Service (IRS) filings from foundations (searchable by organization name, geography, Employer Identification Number (EIN)). It also features free and fee-based online classes, reference guides (under "Get Started" and then "Get Answers") with items of interest to individual grant-seekers and nonprofits, online tutorials, links to nonprofit resources, and much more.

GovBenefits.gov
www.govbenefits.gov/

This site enables individuals to locate Federal and State benefits including loans and grants (over 1,000 programs included) for which they may be eligible. Users can search by keyword or in predetermined topic areas. The site offers a confidential questionnaire to produce a list of programs for which a person may be eligible.

Grants.gov
www.grants.gov

A place for organizations (government, educations, public housing, nonprofit, some small businesses) to find and apply for Federal grants. It lists information on over 1,000 grant programs. The site can perform basic keyword search, browse by category or agency, or perform an advanced search.

International Association of Fire Fighters (IAFF)
www.iaff.org/grants/

The IAFF Grants Administration has staff dedicated to assisting its local affiliate departments with guidance for grant-application writing and review. Web resources include links to all AFG grant applications and guidance.

NVFC Fire and Emergency Services Grant Writing Guide
www.nvfc.org/files/documents/Grant_Writing_Guide.pdf

The National Volunteer Fire Council (NVFC) developed this simple-to-follow guide to help departments navigate the complexities of applying for Federal grants. This guide is designed to make the process less daunting and assist departments and their grant writers in preparing a competitive application.

Proposal Writing Short Course
http://foundationcenter.org/getstarted/tutorials/shortcourse/index.html

Located on the Foundation Center website, this site offers advice on preparing a grant proposal.

Sample Grant Proposals
www.theideabank.com/onlinecourse/samplegrant.php

This website includes sample proposals and narratives from successful grant writers from the AFG program.

Chapter 4: Local Revenue and Funding Alternatives

There are a wide variety of funding mechanisms available to municipalities to meet the cost of public services through both a Budget Plan and a Capital Improvement Program. Facilities and services that are provided to all citizens as a benefit of living in the city are paid for by a number of revenue sources. Rules regarding these revenue sources will vary from State-to-State and agencies using the same funding mechanisms may have very different methods of implementation, approval, or authority. This chapter provides a general overview of various local revenue sources; debt financing (borrowing); emergency medical services (EMS) billing; cost-saving strategies; and volunteer and nonprofit fundraising ideas.

Sources of Local Revenue

The most common source of funding for local governments is taxes. These include property taxes, sales tax, excise tax, income tax, and an assortment of regulatory and user fees. Which types of taxes available to local governments is determined by each State Constitution? Senior executives in emergency services must familiarize themselves with how their jurisdiction is funded and understand the political and legal limitations of taxes as a source of local revenue.

Taxes

Cities, counties, and special districts, such as a fire district, have the authority to levy a variety of taxes. These taxes are either general purpose or special taxes. Special taxes must be used for the specific purpose for which they are imposed. The word "tax" has become one of the most dreaded words in the American lexicon. The attitudes and preferences of those who bear the tax ultimately will determine the tax structure in the local community.

Local Property (Ad Valorem) Tax

An ad valorem tax (Latin for according to value) is a tax based on the value of real estate or personal property. It is perhaps the most common source of funding for municipal and county services. A property tax is typically levied at a set rate per dollar of assessed value. There are two forms of property tax: primary and secondary. The primary component of a property tax is used to fund general operating expenses, while the secondary component is used to fund special obligations, such as the repayment of bonds and budget overrides.

Property taxes may provide an advantage for local governments in that they are 1) a potentially large and stable source of revenue, 2) are exportable to absentee property owners who escape local sales or income taxes yet still benefit from public safety services, 3) are familiar to citizens and business owners, 4) have limited risk of taxpayer avoidance, 5) are equitable in that the benefits derived from the services funded by the tax are associated with higher property values, and 6) they are deductible from Federal income taxes.

While the property tax remains the mainstay of the local revenue structure, it is one of the most politically-unpopular taxes. A key reason for dissatisfaction is that a property tax falls on unrealized gains in property values, making it punitive for those who may be property rich but cash poor, such as senior citizens who live on fixed-retirement incomes. Many States have experienced taxpayer revolts that limit increases in property valuation or require a super-majority of voters to approve any tax increases. The most well-known examples include Proposition 13 in California and the Taxpayer Bill of Rights (TABOR) legislation in Colorado.

States that limit or prohibit a local property tax may authorize a "parcel tax" instead. Real estate is divided and sold in parcels. A parcel tax is similar to a property tax in that it is a tax on land. It differs in that the parcel tax is a tax on the real estate parcel itself, not on the value of the parcel. Parcel taxes are commonly used in California to fund school districts and fire and EMS districts.

Administrators considering property-tax increases to fund fire and EMS needs should consider implementing remedies to reducing taxpayer dissatisfaction such as tax-deferrals to provide tax relief to low-income residents; Homestead programs that exempt a fixed dollar-amount or percentage of a property's assessed value; or split tax rolls that tax single-family residential properties at one rate and all other properties at a higher rate.

Taxpayers may also be more agreeable to a property tax when its revenues are dedicated solely to fire and EMS. For example, in May 2007, voters in Queen Creek, AZ, approved that community's first property tax dedicated to the establishment of a public fire and EMS department.

Fire Flow Tax

The fire flow tax is a type of property tax that is assessed to properties based on a computed fire flow requirement, typically using an Insurance Services Office (ISO) formula for fire flow. The tax can be used to cover the cost of fire protection and other emergency service functions. The fire flow tax amount is determined by calculating the risk factor of a property based on a specific formula. The Morgan-Orinda (CA) Fire District charges a rate of $0.06 per unit of risk to determine the tax bill for each property.

An advantage of the fire flow tax is that it can generate significant revenue and the charge computation can be computerized and done automatically. A fire flow tax can also be used to incentivize fixed fire protection systems, such as residential fire sprinklers. The Moraga-Orinda Fire District allows a reduction of 50 percent for residential fire sprinklers. Since the fire flow tax is due annually, the benefit of the reduced tax recurs each year.

However, the fire flow tax is a tax and faces many of the same issues as traditional forms of taxation, including opposition from taxpayers and businesses.

Sales Taxes

Next to property tax, sales taxes are the most important revenue sources for local governments. A sales tax generates revenues by imposing a tax on retail and other sales activities. These taxes go into a community's general revenues that support a myriad of services including fire protection and EMS.

A sales tax tends to be a more popular choice for raising revenue than a property tax. An important reason for this is that the tax is paid only when someone purchases an item. In this way, the tax is collected in small increments over time. The tax also reaches nonresidents who shop or visit a community and consume municipal services, but do not pay a property tax.

It is not uncommon for sales taxes on some goods to be earmarked for purposes benefiting consumers of those goods. An excise tax is a type of sales tax that is applied to selective products or services. Excise taxes are typically benefits-based in that they are intended to recover at least part of a public service from those who benefit from it. In 2009, the New Mexico Legislature enacted the EMS Rescue Act, which increased the Liquor Excise Tax to improve delivery of EMS, trauma, stroke, and cardiovascular emergency services.

Communities that have a thriving tourism industry often enact excise taxes on services that cater to tourists and conventioneers. Such taxes are sometimes referred to as a "transient occupancy tax," or bed tax. Visitors often place a large demand on emergency providers, but may not directly or fully support the services they use through property taxes on hotels, restaurants, or other tourist industries. Various excise taxes help make up the difference. For example, a bed tax may be added to the cost of a hotel room and additional taxes tacked on to the cost of a car rental. The revenue generated from these excise taxes typically go to a city's general fund which pay for a variety of services.

Excise taxes are especially appropriate to consider where the transient population is large relative to the resident population, or where the transient population places a disproportionate demand on emergency services.

As a rule, the excise tax on tourist industries typically enjoy high political appeal because of their relative ease to administer locally, and because the burden of the tax is on nonresidents, not local voters. They do, however, muster significant opposition from local tourism industries and a city's Chamber of Commerce.

One consideration in implementing a transient excise tax is the local tax burden in comparison to neighboring communities. Increased taxes could decrease demands for those services and offset any new revenues expected from the tax. If the tax rate is noticeably higher in a community, business revenues and tax collections may be lower as visitors stay in neighboring jurisdictions with lower tax rates. This possibility must be evaluated in the context of the jurisdiction based upon the type and number of visitors, local economic conditions, and the tax rate of neighboring, competing communities.

Sales taxes are subject to certain exemptions, particularly for goods and services considered "basic needs." This is typically done to balance out the regressive nature of the sales tax. The most common exemption of this type is for grocery food items and prescription drugs.

There are growing concerns about an overreliance on sales-tax revenue. Over time, households are spending more money on services, which are generally not taxed. Another problem is the growth of Internet sales of goods is eating away at the sales-tax base. Virtual or Internet companies that do not have a physical presence in a community but ship goods for sale to a buyer within that community may be beyond the ability of the jurisdiction to collect.

Sales-tax revenue is subject to the ups and downs of the economy. During the "Great Recession," significant declines in sales-tax revenue in many parts of the country led to local governments making blanket budget cuts, brownouts, layoffs, hiring freezes, labor concessions, and benefit-cost reductions.

Income Tax

An income tax is typically assessed on the wages and earnings of individuals, but may also be applied to the net income of unincorporated small businesses. Forty-three States levy some level of income tax. In a few States, local governments can also levy an income tax. One of the strengths of a personal income tax is the capacity to reach nonresidents who commute to jobs in the city. These commuters use city services but do not contribute their fair share for the cost of providing these services through property or sales taxes.

A portion of the City of Delaware, OH, income tax is dedicated to fund fire and EMS in that community. In November 2010, citizens voted to increase the levy 3/10 of 1 percent to provide additional funding for the city's emergency services. The new revenue was used to make capital improvements to aging fire stations and add capacity to the city's growing southeast side.

Income-tax funding for public safety may also come in the form of surtax. Surtax is an add-on tax that is assessed as a percentage of the Statewide income tax owed. For example, if a taxpayer owes $1,000 in State income tax and if the surtax rate is 2 percent, the amount owed in income surtax to the local Fire or EMS District would be $20.00. Counties in Iowa may impose a countywide EMS income surtax, though few actually do.

Like property taxes, State and local income taxes are deductible from the Federal income tax. Another benefit is that the income tax is based on the ability to pay. However, since a large number of small businesses pay a personal income tax, Chambers of Commerce historically oppose the assessment of local income taxes. Income taxes are also sensitive to the business cycle and are not as stable as property taxes.

Real Estate Transfer Tax

Real estate transfer taxes are special-purpose taxes assessed on the sale of property. Usually, they are a percentage of the selling price of the real estate. Real estate transfer taxes have sometimes been levied to provide an additional source of revenue for public safety and public works projects. Proceeds from such taxes are pooled with other general-fund revenues but can be earmarked for specific purposes.

Thirty-five States and the District of Columbia impose some form of real estate transfer tax. California, Louisiana, and Ohio real estate taxes are imposed only at the local level. In Delaware, Maryland, Michigan, New Jersey, Pennsylvania, Washington, and West Virginia, some localities may impose a real estate transfer tax in addition to the State transfer tax.[3]

Most often, these taxes are used in areas with high single-family home ownership. Unlike property taxes, which are passed on to renters and low-income residents, a transfer tax is imposed only on those with incomes sufficient to purchase real estate. In some cases, first-time homebuyers may be exempted from the tax. Another advantage is that they are easy to collect so they have a low administrative cost. They can be collected along with property taxes at the time of closing on the mortgage or when the deed transfer is registered.

However, real estate transfer taxes are not without controversy. The tax is heavily opposed by realtors, home builders, and other real estate interests. Arizona preemptively banned local governments from implementing them and other States and localities have considered repealing the tax. Since most States require a local jurisdiction to get legislative approval to implement a real estate transfer tax, adoption is almost certain to face significant political opposition.

Utility-User Tax

A utility tax is a charge on the use of public utilities such as telephone, cell phone, cable television, gas and electric services, municipal water, wastewater, and garbage collection. The utility tax applies to both businesses and homeowners. Taxes are collected by the utility as part of its regular billing procedure and then remitted to the city. A utility-user tax may be imposed as a special tax, earmarked for a specific purpose, or a general tax to be used for a variety of municipal needs.

Proceeds from the utility-user tax are used to fund local government services. The tax pays for law enforcement, fire protection, EMS, maintenance of city parks and streets, youth programs, and other general-fund services. Laws may include exemptions for seniors and low-income residents.

The Western Wayne County Ambulance Trust Authority covers the communities of Stillwater, Perkins, and Glencoe, OK. In 2011, the Authority implemented a Resident Benefit Program attaching a $5-per-month fee to residents' utility bills. The fee covers the utility account holder and all permanent members of the household. Residents can opt out of the program but are responsible for the full cost associated with prehospital medical treatment and transportation.

Development Impact and User Fees

Development impact and user fees are imposed to pay for the cost of programs or facilities that reduce the negative impact of an activity or specific business on a community. Fees are charged to ensure that those benefiting from an activity pay their fair share of the costs related to that activity. Costs may include licensing or permitting, cost recovery, inspection, and enforcement costs. Examples include an Enhanced 9-1-1 (E9-1-1) fee assessed against local communication services to recover costs for services and equipment that allow customers to dial 9-1-1 emergency services.

3 Federation of Tax Administrators: www.taxadmin.org

Development Impact Fees

An impact fee is a direct charge levied by local governments against developers to help offset the cost of new growth. Impact fees most often take the form of a one-time permit charge assessed at the time of plat approval or an application for a building permit. These fees provide a city funding for capital projects. Cities may only impose fees on developments that will benefit from the infrastructure improvements. The fees cannot be used to fund operational expenses. Therefore, impact fees cannot be used for maintenance or to eliminate deficiencies in older neighborhoods.

Impact fees can provide some financial relief for growth-related problems. In California, some cities collect fees for such urban-growth needs as road improvements, mass transit, public art, low-income housing, day-care centers, and job training. Arizona allows impact fees for fire, police, parks, recreation, libraries, public buildings, and streets. Like real estate transfer taxes, impact fees are facing increasing political scrutiny from real estate interest groups. Today, 26 States have implemented the use of impact fees, mostly in the western States, along the Atlantic coast, and in the Great Lakes region of the country.

User Fees

Cities have the authority to impose direct charges, or fees, on individual users of services. Use of these revenues is restricted to paying for the service for which the fees were collected. User fees are a fairly efficient way to distribute the costs of government services. Many communities are serviced by private ambulance providers. As such, a price can be affixed to some services, such as transportation or EMS-standby services, and the customer billed for the provision of those services. Fire departments and EMS agencies have assessed a number of EMS-related user fees as a means of cost recovery or alternative funding.

Emergency-Response Service Fees

Fire and EMS agencies have experimented with charging fees to insurance companies to raise revenue to support services. Typically, automobile insurance policies provide coverage for medical expenses and ambulance transportation, but not for fire- or police-response services. These fees try to recoup the cost of providing noncompensated prehospital medical treatment and rescue activities.

This fee is not without controversy. Proponents point out that a high portion of motor-vehicle accidents to which fire and EMS agencies respond to involve drivers who are nonresidents and not part of the local tax base. Opponents of the fee, particularly the insurance industry which calls it a "crash tax," claim that emergency responses to vehicle accidents are part of the regular duties of first responders and are funded by local taxes. Nonresidents will pay sales taxes and transient taxes that help to cover their portion of the costs. Arizona, Utah, and Kansas have preemptively banned emergency-response service fees.

Inspection Fees

Inspection fees have long been used by fire departments to provide funding for fire prevention. Fee schedules vary among jurisdictions. Inspection fees may be based upon the type of inspection conducted (initial or reinspection), the occupancy (educational, industrial, residential, etc.), and the size of the building. Many departments charge a flat fee for initial inspections with additional fees for each subsequent reinspection. Additional fees may be charged when special hazards are present, such as hazardous materials storage areas.

Business self-inspection programs have become an accepted way to address low-hazard occupancies, while making certain the fire department has the necessary information on businesses operating within the jurisdiction and the types of hazards present. Self-inspection programs involve sending a checklist to the registered business operator. The owner completes and returns the checklist to the fire department along with a nominal fee in exchange for a Certificate of Inspection. Some fire departments conduct random audit inspections to ensure proper compliance. Self-inspection programs only work for low-hazard occupancies. The advantage is that self-inspections collect information the firefighters need, while also providing a revenue stream for fire prevention services at little cost to the department. The following is one example of a fire inspection schedule.

FIRE INSPECTION FEE SCHEDULE

Effective From June 15, 2010 to June 15, 2011

Existing Fire Prevention Inspection Fees

(a) A fee shall be imposed by the Village for each inspection performed by the Fire/Rescue Department and shall be according to the fire inspection fee schedule herein set forth below.

ANNUAL FIRE INSPECTION FEE SCHEDULE		
Assembly Occupancies (Based on Occupant load)		
A-3	Class A (greater than 999)	$162.98
A-2	Class B (between 300 and 999)	$114.09
A-1	Class C (between 50 and 299)	$70.63
Educational		
E-1	Under 5,000 sq. ft.	$54.34
E-2	5,001 to 10,000 sq. ft.	$108.66
E-3	All others	$54.85
	Daycare centers	$54.85
Health Care/Institutional		
	Ambulatory health-care centers	$54.85
	Limited-care facilities	$54.85
C-1	Nursing homes	$162.98
	Hospitals	$271.64
Detention and Correction Occupancies		
	All types	$271.64
Residential Occupancies (per unit charge)		
M-F	Multifamily 1-2 stories	$5.44
M-F	Multifamily 3-4 stories	$5.44
M-F	Multifamily 5 stories and over	$5.44
A L	Assisted living (per bed)	$5.44
D-1	Hotel or motel facility (per bed)	$5.44
Mercantile, Business and Storage		
B-1	3,000 sq. ft. and under	$54.85
B-2	3,001 to 6,000 sq. ft.	$70.63
B-3	6,001 to 10,000 sq. ft.	$114.09
B-4	10,001 sq. ft. and over	$162.98
Industrial/Manufacturing		
F-1	Under 12,000 sq. ft.	$81.48
F-2	12,000 and over	$162.98
Other structures and required permits by National Fire Protection Association (NFPA)		$54.85
Review of fire and/or disaster operational plans		$54.85
Locked or blocked exit door will be immediate fine of		$81.59 each

Initial fee covers initial inspection and one reinspection trip. Final inspection fee will be based upon the number of inspections needed for compliance, in accordance with reinspection fee schedule.

REINSPECTION FEE SCHEDULE	
All additional inspection trips shall have accumulative charges based upon the following schedule.	
	Total charge
Single inspection trip	AFIFS*
Second inspection trip is no charge	$0.00 + AFIFS*
Third inspection trip fee is $38.04	$38.04 + AFIFS*
Forth inspection trip fee is $76.08	$144.12 + AFIFS*
Fifth inspection trip fee is $$152.16	$266.28 + AFIFS*
Sixth inspection trip fee is $304.32	$570.60 + AFIFS*
Seventh inspection trip fee is $608.64	$1,179.24 + AFIFS*
Eighth inspection trip fee is $1,217.28	$2,396.52 + AFIFS*
Ninth inspection trip fee is $2,434.56	$4,831.08 + AFIFS*
Tenth inspection trip fee is $4,869.12	$9,700.21 + AFIFS*
Each additional trip doubles the previous inspection charge.	

*AFIFS is the Annual Fire Inspection Fee Schedule.

As of 4/29/2009, per Chief, anything other than listed specifically above gets charged minimum Plan Review of $80.72.

(b) A fee shall be imposed by the Fire/Rescue Department for the review of plans, drawings, specifications, engineered submittals, shop drawings, sketches, and for inspections for all new construction, renovations, or demolition within the Village and shall be according to the fee schedule below.

FIRE/RESCUE PLANS REVIEW AND CONSTRUCTION INSPECTION FEE SCHEDULE

The following formula, which is based on the square footage of the proposed work, will be used to determine Plan Review fees.

$0.27 for every square foot of construction, demolition, and renovation of construction in the Village of Tequesta, except for single-family residential occupancies.

Minimum plan review and permit fee shall be $81.48.

Review of plans or specifications that are not sealed by a licensed architect or engineer shall increase the permit fee by 50 percent.

Commencement of work prior to issuance of permit shall double the required permit fee.

A fee of $38.04 shall be charged for each failed inspection, which requires an additional trip to the job site.

Plan Revisions

Minor revisions:	10 percent of original fee
Major revisions:	50 percent of original fee
Minimum revision fee:	$54.34
Restamp:	$5.43 per page

A fee of $108.66 shall be charged for a waterflow test. Fee to be paid in advance.

The Department of Fire/Rescue services shall charge $92.36 per hour for all consultations such as prepermit meetings, bidding conferences, conceptual design reviews, Development Review Committee (DRC) review, and/or conferences. The minimum charge shall be $135.82.

Required permits under the Florida Fire Prevention Code which are not included in the construction square footage fee shall be in accordance with the following table. Minimum fee of $54.34, plus the following charges.

Fire Suppression Sprinkler System (charge per head)	$1.09
Liquefied Petroleum (LP) Gas Installation (charge per connection)	$2.18
Fire Pump installation and acceptance test	$108.66
Fire-Suppression Hood	Minimum Fee
Hood suppression system (per nozzle)	$2.18
Standpipe system (per hose outlet)	$5.44
Fire Alarm System (per device)	$1.09
Other required permits (includes shutters)	Minimum Fee

Fire watch details that are required by the Code and/or required by the fire chief shall be charged at a rate of $38.04 per hour.

Commencing May 2007, and every year thereafter, all hourly rate charges and permit fees shall be adjusted annually by an annual cost index. The annual cost index shall be the average of the May Consumer Price Index for All Urban Consumers, Southern Region, and the May Consumer Price Index, Medical Care Group.

Plan Review and Permitting

Many fire departments review building plans for fire code compliance and inspect the installation of fire protection systems during construction. The fire department often receives part of the permit fees paid to the jurisdiction for these services.

Fees are also charged for occupancy permits, special hazards permits, reviewing plans for building renovations, and reviewing new fire protection systems in existing buildings. Fees are also often charged for inspecting daycare centers, hotels, hospitals, nursing homes, spray-painting businesses, and other specific occupancies that require special permits to operate.

Departments may also charge a fee for special event permits such as public events, the use of fireworks, large tent events (circuses, beer tents, etc.), as well as other special purposes such as open burning or movie production sets.

Often, the problem with plan review and permitting is that the fee is not set high enough to recover the full costs of the services provided. They are among the fees that developers, builders, and others expect to pay but object to any increases in the fees. Local governments need to consider whether they want the fees to cover the entire cost of providing the service or only a portion, and whether they want them indexed in some way or recomputed annually or at least every few years. These are policy decisions made by elected officials and need the informed input from the fire chief and fire marshal. Departments may encourage waiving all or part of these fees to encourage installation of additional automated fire protection systems.

Hazardous Materials Fees

Maintaining the capability to respond effectively to hazardous materials incidents adds significant costs for local jurisdictions. Hazardous materials response requires hundreds of hours of training and continuing education, specialized equipment, and technical expertise to conduct inspections. Hazardous materials occupancies do not have to be large or unusual to pose a challenge, such as a microchip manufacturing plant, the local pool store and exterminator business can pose significant problems for first responders. Seemingly minor incidents involving hazardous materials can keep fire companies occupied for long periods of time and present dangers to the public, responders, and the environment.

To offset the expense of providing hazardous materials response capabilities, some fire departments have adopted a hazardous materials storage and inspection fee. Revenue from this fee helps ensure steady income for training fire inspectors and covering the cost of specialized inspection services.

Many fire departments also charge for hazardous materials team response, both to offset the cost of the response and to incentivize proper maintenance of hazardous materials facilities. The fee also helps replace equipment used to mitigate a spill or release. Federal law requires the owner or transporter of released hazardous materials to pay cleanup costs, including fire department and EMS costs, which helps to justify these fees.

Special Service (Standby and Fire Watch) Fees

Fees for "special" services attempt to recover or offset the costs from the users of the service. These fees may be charged for services such as EMS standby at a football game or fire watch at a concert venue. The users often pay less than they would if they contracted with a for-profit provider for the service and often receive intangible benefits such as communications links that can quickly get additional resources to an event if an emergency develops.

Sometimes, fire and EMS agencies provide personnel to who serve offduty and are paid directly by the special event, rather than through the agency or local government. Either way (onduty or offduty), protection is provided for a special purpose, and the agency receives reimbursement for the service.

The downside of these special fees (and fees in general) is that the public may expect these services to be provided routinely, without further charge, as part of their taxes. A public education campaign may be necessary to prevent discontent and resistance. Special services fees can be presented to the public as improving equity to all taxpayers, by not using everyone's taxes to subsidize those who require services above the level provided to everyone else.

In addition to stand-by fees, fire departments may charge special service fees for such services as hazardous materials response, water rescue, fire protection system resets, fire inspections, and permitting.

Emergency Medical Services User Fees

Many fire departments have historically offered EMS without charging a service fee, unless the patient was transported to the hospital. Increasingly, fire and EMS agencies are implementing nontransportation-related fees for service in an effort to recoup cost of supplies. The San Antonio Fire Department charges whenever EMS personnel are called out to do a medical assessment, even if the patient isn't transported to the hospital. This nontransport charge averages $85. The department also charges a fee for EMS supplies and medications ranging from $5 for Albuterol or Dextrose to $170 for Diazepam and Amiodarone.[4]

The EMS service fee has been successful in paying for a portion of the cost of maintaining EMS capacity. Similar fees have helped fire departments maintain existing levels of service and, in some cases, engage in new services.

Fines, Forfeitures, and Penalties

In some U.S. cities (and several other Nations), people responsible for fires or certain high-risk rescue activities may be held financially accountable for their actions, and to pay for the cost of emergency services. Recovery of costs is intended to not only defray the cost of providing rescue services, but to change behavior in order to deter risk-taking behavior among the public.

Depending on State law, cost-recovery measures may need to be approved by legislative act. Taking people to court to recover costs, such as suing after a large incident, are often unsuccessful without previously enacted legislation.

For example, Section 28-910 of the Arizona Revised Statutes authorizes fire and rescue organizations to bill motorists who knowingly cross a barricaded flood area. Known as the "Stupid Motorist Law," the law allows fire departments and rescue organizations to impose a fee (or fine) for the rescue of a motorist out of a flooded area, up to a maximum of $2,000.

Alaska adopted legislation to assist in enforcing seatbelt use as well as to generate revenue for EMS agencies in the State. The legislation states that if a person is guilty of an infraction concerning seatbelt usage, they can be fined up to $15 per person. Fines are higher (up to $500) for children not in proper Alaska safety-restraint systems. The courts have the ability to waive this fee if the person convicted donates $15 to the EMS agency providing services in the area which the violation occurred. This program generates supplemental funding for EMS agencies in the State.

The Ohio EMS grant program funds improvements to enhance EMS and trauma patient care in the State. All EMS and trauma system grants are funded through the collection of seatbelt fines.

California Senate Bill 12 established the Maddy Emergency Medical Service Fund. Maddy is a mandated program funded by revenue generated from fines, forfeitures, and penalties collected for all criminal offenses. Funds are intended to reimburse providers for the costs associated with providing emergency care

4 San Antonio Fire Department EMS Billing Policies and Fees: www.sanantonio.gov/safd/emsbilling.asp?res=1024&ver=true#listing

to uninsured patients. Seventeen percent of Maddy funds in Merced County support the county EMS agency that provides oversight of EMS, such as certification of emergency medical technicians (EMTs), monitoring ambulance services, and expenses related to the maintenance of quality emergency-response systems.

Fines for Nuisance Alarms

Most new commercial buildings and an increasing number of residences have fire detection systems that can trigger unwanted fire alarms requiring the response of the fire department. Each false alarm creates some danger for the public and firefighters. In 2009, fire departments responded to 16 false alarms for every 10 fires, and 45 false alarms for every 10 structure fires.[5] From 2000 to 2009, 24 firefighters died responding to unwanted fire alarms, including malicious false alarms and alarm malfunctions.[6]

Increasingly, cities are adopting nuisance and unwanted alarm ordinances that include fines to encourage better maintenance of systems, place greater responsibility on the system owner for unnecessary or inappropriate actions triggering alarms, and recover some of the costs of responding to these types of alarms.

According to the NFPA, one-quarter of jurisdictions issue violation notices for false alarms and almost one-third (31 percent) assess fines or penalties. This approach has proven effective at reducing nuisance alarms. A March 2007 NFPA Journal article entitled "Nuisance Alarms" documented a successful management approach to addressing a false alarm problem in three multitowered commercial highrise buildings in Calgary, Alberta, Canada. Through a combination of increased training, improved procedures and communications, investigation of causes, and assessment of fines and penalties, false alarms were reduced by 50 percent between 2003 and 2006.

The False Alarm Reduction Association (FARA) has a model fire alarm ordinance available on their website (www.faraonline.org).

False medical alarms are increasingly becoming a problem for fire departments and EMS agencies and some cities have instituted penalties for these incidents as well. Rockford, IL began charging a fine for false medical alarms in February 2010. A fine is assessed for more than five false alarms per year. The primary subscriber of the medical-alarm service is assessed a $100 fine for each false alarm from 5 to 8, $200 for the 9th and 10th incident, and $300 for each false-alarm response above that.

Seized Assets

Another source of funds, equipment, and vehicles accessed by a growing number of public safety agencies are the assets seized during drug raids. Where the fire and EMS agency can demonstrate that illegal drug activity has increased the demand for services, such as through EMS records of illegal drug overdoses, or that first responders have participated in drug-related incidents (such as hazardous materials team response to drug labs, Special Weapons and Tactics (SWAT) paramedics in support of drug raids, and treating victims of raids), they may be able to share in money and equipment seized by law enforcement in drug-related arrests and raids. They money may have to be used for the purchase of special equipment for assistance in drug-related incidents, but can include vehicles, ambulances, communications equipment, computers, and other resources. The equipment does not have to be used solely for drug-related incidents so long as it is available for such incidents.

---

5 "Unwanted Fire Alarms." NFPA, 2011.

6 "Firefighter Fatalities in the U.S." NFPA, 2010.

Enterprise Funds and Utility Rates

The enterprise fund consists of a city's utility programs such as electric, natural gas, water, wastewater, and trash collection. Utility rates are charged to users of a city-provided service. The net income from these utilities can be transferred to the general fund to help pay for operating costs for general governmental services such as police, fire, EMS, and parks and recreation.

Often, local governments establish an enterprise fund for city-operated ambulance service. For example, the Carson City Fire Department has an ambulance enterprise fund which bills a patient's insurance company for ambulance transportation and related prehospital-care expenses. The revenue generated from these services pays for the operation and maintenance costs of the ambulance services, including paramedic salaries. The City of Petaluma, CA operates its ambulance service as an enterprise fund with the fees charged for service covering all expenditures of the program.

Sale of Assets and Services

The sale of surplus department assets or charging for special services can provide a new revenue stream.

Sale of Used Equipment

Some departments have raised revenues by selling used equipment and apparatus to other agencies or to collectors. The Wisconsin EMS Association maintains a webpage that contains a listing of used EMS equipment placed for sale by nonretail organizational and individuals.[7] Listings range from ambulances to gurneys, cascade systems, and cardiac defibrillators/monitors.

Sale of Services

Some EMS agencies sell services to other organizations. A principle example of this is charging fees for training to other local agencies or the private sector. These charges may just offset costs, or they can be established to provide net income. Training services are provided either as a fee-per-student or on a contract basis.

When considering the use of training as a source of revenue, several factors should be considered to ensure that the training of others does not negatively affect the training of the fire department or EMS agency's own personnel. Many departments are cash-strapped or unable to provide all of the diverse EMS continuing education required today. The department offering to train others must make sure it has the qualified personnel and facilities available to provide the training without overloading its instructors or interfering with its scheduled training. Serious consideration should also be given to liability.

In addition to providing training services to other organizations, many departments provide training to private companies and individuals for a fee. Training such as cardiopulmonary resuscitation (CPR) courses, babysitter classes, and first-aid certification require relatively few resources and fit well within the mission of the department. They can be quite profitable, too.

City fire departments have also been involved in selling driver-training services to industry; providing medical physicals for personnel in smaller departments; providing mechanic services; and contracting EMS billing services. For example, the Carson City, NV, Fire Department, Emergency Medical Services Division provides ambulance billing service for the Central Lyon County Fire District as a means of generating additional revenues.

Sale of Delinquent Tax Certificates

Local governments rely on property taxes to fund essential services such as public schools, law enforcement, road construction, parks, and fire and EMS services. When taxes are not paid, the government does

7 Wisconsin used EMS equipment website: www.wisconsinems.com/index.php?module=cms&page=27

not have the operating capital it needs to fund these programs. As a means of recovering lost income from delinquent taxpayers, county governments offer tax-lien sales at auction to the public. The purchaser pays the delinquent taxes on behalf of the property owner and in exchange is given first-lien position on the title, ahead of mortgages, deeds of trust, and judgments. Under the terms of the sale, which differ from county to county, if the debt is not repaid in a specified period of time, the purchaser of the tax lien may foreclose on the property.

Facility Rentals

Fire departments often rent out meeting space for private functions. Some departments intentionally design new stations to include meeting space that is not only useful to the department, but also serves as a community facility. Departments can rent out a dining hall or large room for dances, parties, exercise classes, weddings, and other gatherings. Some of these organizations provide catering upon request. The fire department needs to ensure that they have adequate liability insurance for hosting these types of events.

The Mountain Preserve Reception Center is owned and operated by the United Phoenix Fire Fighters Local 493. The Mountain Preserve offers a view of the Valley and is available for conferences, banquets, and weddings. Conference rooms provide meeting space for union events and meetings and are rented out to generate revenue for the organization.[8]

Facility rentals may face the same type of opposition as fees for service in that the public may feel they are being double-charged for a facility—once through taxes and again with a user fee. As with other charges, a department must weigh the potential public indignation in charging for service against the revenue raised. Charging for events that are clearly private affairs may prove more acceptable to the public.

Fire Protection Subscription Service

Used more frequently in EMS, subscription services for fire protection may be offered to residents or property owners outside of an incorporated city or town or organized fire district. A contract is set for a specified period of time and the fees based on a cost-per-square-foot or risk-based formula.

As of 2009, Karns, TN, Volunteer Fire Department (VFD) received no tax funding for its services and only 21 percent of community residents donated to the agency. In 2010, the Karns VFD successfully implemented a subscription service program asking residents to pay $0.07 per square foot of property. The average subscription fee is between $100 and $200 per household. The department responds to all requests for assistance, regardless of membership. However, nonsubscribers are billed for services including fires, automobile accidents, gas leaks, fire alarm activations, illegal burns, and any other type of call. Fees for responding to a structure fire start at $1,900 for the first hour. Revenue raised from the subscription program is being programmed for apparatus purchases and the construction of a new fire station. Karns' firefighters remain volunteers despite the change in funding method.

Subscription programs have created controversy when fire departments refused to extinguish fires in nonsubscribing properties. This policy also does not fare well with community residents and voters. Communities considering a subscription service program need to clearly inform the public about the planned rules, fees, and penalties in order to reduce misconceptions.

Benefit-Assessment Districts

Benefit assessments are a common financing tool. Assessments are charges on real property or businesses levied to pay for facilities or services within a predetermined area according to the benefit resulting from

8 Mountain Preserve Reception Center: www.mprc.net/index.html

the improvements. Typically, citizens or local governments will establish a special district for the purpose of levying an assessment to finance capital improvements or provide local services. There are two types of assessment districts: Special-Purpose Districts and Benefit-Assessment Districts.

Special-Purpose Districts

Special-purpose districts, sometimes referred to as special-district government, are defined by the U.S. Census Bureau as limited purpose governmental units that exist as separate corporate entities that have fiscal and administrative independence from general-purpose governments such as county, city, or township governments. Fiscal independence means that the special district may determine its own budget, levy taxes, collect charges for services, and issue debt without review from another local government. Administrative independence comes from the fact that members of the governing board are elected by the people of the district. Special-purpose districts provide a specific service to residents that are not provided by the general-purpose government. Examples of special-purpose districts include fire protection districts, ambulance service districts, county hospital districts, and county health-care and mental health-care districts.

The Apache Junction, AZ, Fire District is a special-purpose district that provides fire and EMS to the City of Apache Junction and several unincorporated communities in Pinal County. Fire Board members are elected independently and are not members of a city council. Board members provide policy oversight for the district and set the tax rate to fund fire protection.

While most districts rely on taxes as a main source of revenue, some special-purpose districts are operated as an enterprise. Enterprise districts are run like a business, charging customers for services and maintaining a positive cashflow. For example, the Jackson Parish Ambulance Service District in Louisiana is an enterprise district operating similar to a private business where services are financed primarily through user charges.

Benefit-Assessment Districts

Occasionally, a county, city, or township government may form a benefit-assessment district within its boundaries by ordinance. Formation of the district is done to address a deficiency in infrastructure or service delivery which falls short of community standards in terms of public safety, health, or welfare. Unlike special-purpose districts, a benefit-assessment district does not have a separate governing board, rather the county Board of Supervisors or city council manages the implementation of services funded through the district. This is because a benefit-assessment district is a funding mechanism, not an implementing authority.

Services are typically financed through property taxes, parcel taxes, or a special taxing formula assessed to properties based on the direct benefit they receive from the district. Examples of factors that have been used to measure benefit include a fire flow factor as measured in gallons per minute (gpm); a response-time factor based on the distance to the closest fire station; a tanker credit computed in rural areas and reflecting the benefit of having a tanker in placed without fire hydrants; an insurance benefit based on the ISO or similar fire protection insurance rating; use or risk classification of the property; and special services such as prefire planning, hazardous materials team, building inspections, heavy-rescue service, among other special benefits.

The assessment may be levied throughout the entire jurisdiction or may be limited to certain areas or zones. In 2010, the Perry, MI City Council established a special-assessment district within the town limits to defray the cost of providing ambulance service. Ambulance services funded, in part, by a $35 per household tax on all parcels in the town.

A benefit assessment can be an important resource for financing fire protection and EMS in a community. It is especially beneficial in States with restrictions on local property taxes, such as in Washington or California. The downside to benefit assessment is that it may be cumbersome to undertake. In addition to

legislative restrictions, fees for each property must be assessed in a fair manner, billed, and collected. The ability to undertake such as project may depend largely on the cooperation of the local general-purpose government's finance and tax collection departments. Citizens must be made aware that the assessment is specifically to maintain or improve emergency services; otherwise, it may be perceived as just another tax measure. In many jurisdictions, property owners may have the right to protest the assessment and ask for a public vote.

Borrowing

Major capital improvements can be funded by borrowing money and repaying it over time. This can be accomplished with bonds, certificates of participation, not-for-profit revenue bond financing, or traditional borrowing.

Bonds

Bonds are a way for a local government to issue debt or, in other words, raise funds. A city issues bonds to support voter-approved projects and then agrees to pay back the bonds with interest. Bond funding is for assets with a long, useful life such as buildings, utility systems, or vehicles. The assets should have the same useful life or longer than the time it takes to repay the bond. For example, if a rescue station has a useful life of 20 years, the time it takes to repay the bond should not exceed 20 years. Bond funds cannot be used for operations, for instance, employee salaries. Bonds usually carry a lower-interest rate than other types of funding; therefore, it is an attractive way of financing.

One of the considerations in determining interest rate is a local government's ability to repay the bonds. This ability is determined by a private assessment company that assigns the city a "bond rating." In order to receive and maintain a high bond rating and thus a low-interest rate, a city needs to have sound financial practices and policies. One example is to keep enough to maintain an adequate fund balance, typically between 8 and 10 percent, in the city's general fund and enterprise fund. If the fund balance falls below this amount, it could cost the community in the form of increased interest rates on borrowing.

There are three main types of bonds:

1. General Obligation (G.O.) Bonds: G.O. Bonds are a common type of municipal bond. They are secured unconditionally by the full faith, credit, and taxing authority of the jurisdiction issuing them. If the taxes levied by the city or town are not sufficient to pay the debt service on the bonds, the issuer is legally obligated to either raise taxes or broaden the tax base to cover the payments.

2. Utility Revenue Bonds: Projects that fall into this category include items related to a city's enterprises such as gas, water, wastewater, and electric. Bonds used for these projects are repaid from revenues received from the city's customers of that particular utility.

3. Excise Tax Bonds: Excise Tax Bonds are repaid by revenues derived from taxation of a particular activity or asset. For example, a bond issue for the construction of a medical facility may be repaid from an excise tax on cigarettes.

Certificates of Participation

Some local governments and special districts are not allowed to issue bonds. An alternative may be a Certificate of Participation (COP). COPs are financing mechanisms similar to G.O. Bonds, but the principal as well as interest are paid along the way, similar to a home mortgage. They are used to fund large or complex financial obligations in a manner similar to leasing but without violating restrictions on debt financing. The borrower sells certificates of participation to investors who agree to provide funds in exchange for repayment of the principle plus semiannual interest payments on a fixed schedule.

For example, a township government wishes to finance the purchase of a new rescue vehicle which costs approximately $180,000, but the town cannot issue bonds. The trustees of the town can hire an investment banker, bond counsel, and trustee to prepare the appropriate documents, offer the COP securities for sale, and manage the disbursement of funds to COP holders. In exchange for the capital to purchase the vehicle, the township trustees agree to make annual appropriations to repay the certificate holders and make semi-annual interest payments. While the obligation remains outstanding, the certificate holders hold title to the vehicle through the COP trustee.

The advantages of a COP include

- They do not violate restrictions on long-term debt financing common among certain forms of government, e.g., townships or imposed by State law.
- They permit public entities to incur a continuing obligation without going into debt.
- In some instances, they are the least costly public-financing mechanism.
- In most cases, they are tax exempt, which attracts a large investor base.
- Usually, they do not require voter approval.

The disadvantages include

- COPs may be more expensive than direct lease or lease/purchase agreements for making relatively small purchases.
- The risk to the lenders that the local government might not appropriate funds every year to finance the annual payment on the obligation generally makes them more expensive (higher interest rate) than G.O. Bonds.
- The cost of issuing COPs is higher than direct leasing. They involve bond counsel, investment bankers, a certificate trustee, rating agency fees, and printing of investment documents. These costs must be added to the amount financed so they can be paid from the sale of proceeds.
- To achieve favorable interest rates, a reserve fund may also be required. This amount must also be added to the amount borrowed.
- The complex legal requirements associated with this type of financing may make it somewhat cumbersome and time consuming to use.

Even though COPs do not require voter approval, care should be taken to ensure that they are broadly supported by the public.

501(c)(3) Revenue Bond Financing

501(c)(3) Revenue Bonds are a low-cost source of tax-exempt financing for capital-improvement projects for qualified not-for-profit organizations. Just like public sector fire and EMS departments, nonprofit agencies may need to purchase or construct a new facility, improve existing facilities, or acquire new equipment to provide community emergency services. A nonprofit fire/EMS agency may be able to finance expenditures on a tax-exempt basis and receive substantial savings through the use of tax-exempt 501(c)(3) bonds.

To issue 501(c)(3) Revenue Bonds, a governmental authority must act as the issue of the bonds, passing its tax-exempt status onto the nonprofit organization. Because the income earned by purchasers of these bonds

is tax exempt from Federal and, in some cases, State taxes, these bonds traditionally carry a lower interest rate and longer terms than conventional bank loans. This saves the nonprofit agency money and gives them better loan terms.

The benefits of 501(c)(3) Revenue Bonds include

- Tax-exempt bonds financing typically offers interest rates that are 20–35 percent less than conventional taxable alternatives.
- Tax-exempt debt may be issued on a long-term fixed-interest rate bases, compared to most taxable debt usually issued for shorter terms at variable interest rates.
- Funds can be used for fixed-asset projects (land, buildings, equipment) that further the tax-exempt mission of the not-for-profit agency.
- Certain 501(c)(3) borrowers may be able to refinance debt and/or finance working capital used to support operations.
- Bond proceeds can be used to pay for eligible costs while the borrower's funds, including donations, can remain invested.

Traditional Loans

Many fire/EMS agencies, particularly independent volunteer organizations, use traditional forms of borrowing money through banks and lending institutions. These loans are also used for capital improvements such as construction of stations, refurbishing existing stations, and vehicle and equipment purchases. Fire/EMS agencies should exercise due diligence in "shopping" for a loan, searching for the best interest rate and loan terms. Some banks and lending institutions will provide lower-interest loans or different loan options for volunteer, service-oriented organizations, particularly if the lending institution is based in the community being served.

Many States have created revolving loan funds to help finance rural fire and EMS projects. In a revolving-loan fund, a loan is made to an agency and, as repayments are made, funds become available for new loans to other agencies. The Arkansas Department of Health manages a Rural Health Services Revolving Fund. The intent of the legislation was to provide resources to help rural communities retain basic medical services, including EMS. The Illinois Finance Authority maintains a Fire Truck Revolving Loan Program to help rural communities finance the purchase or replacement of fire trucks. The program will loan up to $250,000 on a 20-year loan. The State of Pennsylvania maintains a fire service loan program that provides loans at a 2-percent interest rate.

Entrepreneurial Emergency Medical Services

Providing EMS presents an opportunity for managers to think entrepreneurially by exploring ways to sell EMS to individuals or neighboring jurisdictions. In many communities, fire chiefs are writing contracts with private basic life support (BLS) ambulance companies to reimburse the costs of providing fire-based advanced life support (ALS) services. A few are experimenting with innovative service delivery that not only provides a new revenue stream, but also has the opportunity to provide proactive EMS-based public health services to the community.

Billing for Ambulance Services

Whether an ambulance service is large or small, municipal or rural, career or volunteer, billing insurance for patient care and ambulance transportation allows the department to tap a consistent revenue stream to sup-

port EMS operations. Medicare is the largest single payer for ambulance services in the United States. There are lengthy and specific requirements to qualify for reimbursement found in Medicare regulations. Navigating these rules is challenging, but absolutely essential to a successful billing program. Many third-party insurers also use Medicare rates and guidelines as the foundation for their own reimbursement standards.

Before starting an insurance billing program, an EMS provider must be certified by Medicare as a provider. EMS providers must apply for provider status and submit proof that they meet all State standards for ambulance services. Any service operating with exemptions from State EMS standards will be denied reimbursement.

For an ambulance transport to be covered, ambulance services must be deemed medically necessary and reasonable. This is established by the patient's condition and the treatment provided. The prehospital patient-care report is an essential part of establishing medical necessity.

Proper documentation is absolutely essential for accurately and appropriately collecting reimbursement. Poor field documentation is a major impediment to receiving payment from third-party insurers. EMS managers should establish a quality-assurance program to review paperwork. Medical control should provide periodic review of patient-care report and ambulance transport documentation standards as part of EMT refresher courses and paramedic continuing education.

Medicare will only cover transportation to appropriate facilities. Generally, this means the facility must be equipped to provide needed hospital or skilled nursing care for the illness or injury involved. For first responders, this typically means transporting to a hospital or critical-access hospital. EMS agencies providing facility-to-facility transport have additional paperwork requirements.

EMS agencies may receive reimbursement for mileage. As a general rule, only mileage to the nearest appropriate hospital equipped to treat the patient is covered.

A proper understanding of the beneficiary signature rules is critical to proper billing of Medicare; private third-party payers typically require signatures as well. Crews must obtain patient signatures and understand who can sign on the patient's behalf if the patient cannot sign on their own behalf. Health Insurance Portability and Accountability Act of 1996 (HIPPA) law also requires patients be notified of their privacy rights and a signature obtained to confirm they received this information. Consult www.hippa.org for information about necessary components of the signature form.

It is the responsibility of each EMS provider submitting claims to Medicare and third-party insurers to become familiar with the insurance coverage and reimbursement requirements. The Centers for Medicare and Medicaid Services (CMS) Ambulance Billing Guide is a useful overview of Medicare Part B guidelines. The information guide and other manuals are available online (www.medicarehic.com). A good billing company can help the EMS provider avoid the pitfalls and set up a successful billing program.

Private Ambulance Company Contract Agreement

A fire department that provides EMS response, but does not transport patients, may still be able to recover the cost of EMS through contract agreements with ambulance providers. Many communities contract with private, for-profit ambulance companies for BLS ambulance-transportation services, while the fire department provides ALS paramedic response. Under this arrangement, if a fire department paramedic provides ALS services on the BLS transport vehicle, the private ambulance company can charge ALS rates. A contract can be structured so that the private ambulance company pays the fire department the difference between the BLS and ALS rates, as well as reimbursing the fire department for ALS supplies used during patient treatment.

Emergency Medical Services Subscription Programs

A growing number of jurisdictions are implementing EMS subscription programs as an alternative to directly charging users of ambulance-transportation services. An EMS subscription program is a voluntary membership program that is designed to help protect families and businesses financially in the event of a medical emergency while also helping an EMS provider recover at least some of the cost expended in the provision of critical services such as EMS.

There are two basic types of subscription service. The first is a flat yearly fee charged per household, which covers all charges for any EMS service provided. The second variation is a small annual fee that covers only those expenses not paid by medical insurance. The user signs up for the program and authorizes the EMS provider to file reimbursement claims directly with the user's health insurer when services are provided. Most insurance plans do not cover the full cost of ambulance services. With a subscription plan to the local EMS provider, the user would not be charged for uncovered expense. Nonsubscribers would be sent a bill for the remaining balance.

For example, in 2006, the City of Suffolk, VA, began charging for emergency-ambulance transportation as part of the city's EMS Revenue Recovery Program. The city began offering subscription services for an annual fee of $60 per household to defray out-of-pocket expenses, such as health insurance copayments and deductibles. A residential subscription also includes family members who reside in assisted living or nursing facilities located in the city.[9]

The city of Costa Mesa, CA, implemented a subscription program in 2005, charging $36 per household or business. There is also a $12 financial hardship membership fee for those who can demonstrate a need using a specific formula. The city estimated that revenues raised through the selling of EMS subscriptions would minimally be $210,000, depending on the success of the marketing and advertising campaign.

It is difficult to anticipate how many people will actually subscribe to an EMS subscription program, but experience has shown that about 15 percent of households can be expected to participate. Clearly, any move to adopt a subscription program should be preceded with a thorough market analysis to project a subscription rate, program costs, and anticipated revenue.

The overhead costs of a subscription program and potential loss of patient revenues from subscribers who use ambulance service could reduce overall system revenue. Current estimates by EMS agencies using subscription programs is that subscribers use EMS at a rate about double that of nonsubscribers. However, because subscribers constitute a small portion of the base population, subscriber usage is usually offset by nonsubscribers.

It is important to seek legal advice when setting up a subscription program so that requirements of Medicare and Medicaid are met by the approach taken. EMS agencies must also consider the administrative time and marketing costs that will be necessary upon implementation of the program.

Interfacility Transport

Third-party insurance providers may pay the cost of hospital-to-hospital ambulance transportation for qualified patients, along with medically-necessary supplies and services. The revenue generated from these contracts can help offset the cost of providing ambulance-transportation service in the community. Nearly every for-profit ambulance company offers some form of interfacility transportation service. More recently, public sector EMS agencies have begun to do so as well.

---

9 Suffolk EMS subscription service and EMS Revenue Recovery Program: www.suffolk.va.us/fire/ems.html

Beginning in 2008, the Lisbon, WI, Fire Department began offering interfacility transport from Community Memorial Hospital in Menomonee Falls to other hospitals in the Milwaukee, Waukesha, and Madison areas. The Ashland, WI, Fire Department also provides interfacility transports from starting locations within the city of Ashland to medical facilities throughout Wisconsin, Michigan, and Minnesota.

Many hospitals, skilled-care facilities, nursing homes, and health insurers will contract with an EMS provider for ambulance interfacility transportation services. However, for first-responder EMS agencies, these contracts must not detract from the primary mission to provide emergency response to the critically sick and injured with their jurisdiction.

Paramedic Intercept

Paramedic intercept is ALS treatments provided by a paramedic to a patient being transported by another ambulance service. Typically, the paramedic unit meets the ambulance onscene or "intercepts" them en route to the hospital.

Under an Intercept Agreement, when a paramedic provides ALS services on the BLS transport vehicle, the BLS service may charge ALS rates. The paramedic service provider receives the difference between the BLS and the ALS rate, and the BLS ambulance company provides its patients with a higher level of care while reducing its overhead costs. In this way, the BLS agency becomes a payer for ALS services.

Since insurance companies reimburse only for ambulance-transportation services, a nontransport ALS entity cannot bill the insurance directly. If State law prohibits the ambulance company from billing a third-party insurer for its services, then the ALS service will not be able to recover its costs from insurance.

Paramedic/Physician Assistant In-Home Services

Some fire departments are experimenting with innovative ways to reduce 9-1-1 calls and unnecessary ambulance transports by providing in-home patient treatments. The Mesa, AZ, Fire Department conducted a pilot program where a two-person nontransport response vehicle, staffed with a paramedic and a physician's assistant (PA), provided "house calls" to patients with flu-related symptoms or urgent-care calls that could be treated at home.

While EMS providers can only bill third-party payers for transportation services, a PA (or nurse practitioner) can receive reimbursement for patient evaluation and on-scene treatments. A public-private partnership between a hospital and EMS provider could provide a new, innovative service delivery model that generates revenue for both entities. At the time this manual was written, this program was still in the pilot stages in Mesa. However, this case study provides an intriguing look into innovative services that may provide alternative funding to EMS in the near future.

Cost-Saving Strategies

Creative fire officers and EMS managers must look for new tools and new approaches to acquiring the resources necessary to provide quality EMS. There are innovative tools and techniques available that can help reduce costs or stretch the department's budget. A few methods include cost-sharing, purchasing cooperatives, strategic alliances, consolidation, and outsourcing.

Cost-Sharing

One of the most sensible ways to reduce costs is to share costs for facilities and services with nearby jurisdictions. Collaborating with other agencies can leverage funds to produce a program or allow joint-use of equipment or facilities that would otherwise not be economically feasible. Training facilities, communications equipment, and disposable items are often purchased jointly. Some neighboring agencies may share

the costs of building and operating stations to serve both jurisdictions. The difficulty in cost-sharing is coming to an agreement on control of shared resources and determining fair and equitable costs.

Purchasing Cooperatives

Purchasing cooperatives provide a mechanism to achieve cost-reduction for equipment procurement by aggregating demand to get lower prices from suppliers. The North Central EMS Cooperative, for example, is a nonprofit purchasing group, controlled by its members, that enables volume discount on EMS products, supplies, and services through group purchasing. In a little over a decade, the co-op has grown to over 2,000 members across 49 States, Canada, Trinidad, the Bahamas, and the United Arab Emirates. Co-op members have access to volume-discount purchasing with name-brand vendors. Savings range from 10 to 55 percent on an array of equipment and disposable items, free shipping, and special ambulance pricing.[10]

Strategic Alliances

While some local governments join together to make capital purchases or share operating costs, other achieve something similar by forming strategic alliances to provide services to surrounding jurisdictions. This can help receive budget problems in both the jurisdiction providing the service and those receiving the service. It can also provide benefits in the form of increased levels of service through economies of scale and by having more equipment and personnel available to serve either's needs.

Forming strategic alliances may be used for providing basic services, or just for special services such as heavy rescue, hazardous materials, or ALS response.

Strategic alliances most often involve a smaller municipality aligning with a larger neighbor to provide service that they would not be able to afford independently. A small suburb, for example, may not be able to afford to equip and staff its own independent, paid fire or EMS department, but may be able to pay a bordering city or county to provide services for them. The Phoenix Regional Dispatch Center is a strategic alliance of over 20 area fire departments that receive dispatch and communications services from the Phoenix Fire Department on a contract basis. This arrangement cuts the cost of separate dispatch centers and personnel, and improves automatic aid and incident management operations.

Consolidation

The consolidation of several departments into a single agency may provide one of the most efficient and cost-effective means of providing EMS. Savings are realized through the elimination of duplicate functions such as management, training, and communications. Consolidation may be thought of as the ultimate version of cost-sharing, on a larger, more permanent scale. The newly formed Indianapolis Emergency Medical Services (IEMS) was a consolidation of Wishard Health Service and the Indianapolis Fire Department EMS agencies. IEMS is itself part of the new division of the Department of Public Safety which merges the financial strength and business acumen of Health and Hospital Corporation; the city's public safety resources; and the medical expertise, performance, and research assets of the Indiana University School of Medicine.

There are obstacles to consolidation. One of the largest can be a power struggle that often occurs as chief officers and managers are, understandably, reluctant to give up any control over operations. Political bodies and citizens often fear loss of control in getting locked into a level of service and corresponding expenditures for a long time. Unions could resist over issues of compensation, personnel rules, and other issues. Sometimes, the cost-per-capita increases for some parties to a consolidation.

---

10 North Central EMS Cooperative: www.ncemsc.org/

Outsourcing

It is common for businesses and government to contract-out for services to an external provider. Organizations that outsource do so for several reasons including: reducing the cost of the service to the organization; placing more focus on the core business; improving quality; accessing special expertise or technical proficiency; access to new information technology; and transferring liability.

Ambulance billing is a frequent topic of outsourcing discussions in departments that provide such services. Billing Medicare, Medicaid, and private insurance is complex, time-consuming, requires special expertise, and, if done incorrectly, can result in a loss of revenue and even subject the agency to external audits and fines.

When considering outsourcing services, care should be taken to draft a comprehensive Request for Proposals (RFP) to evaluate potential contractors. The RFP should clearly define performance standards, measurable goals, and clearly-defined fee structure. The billing agency should demonstrate expertise with the rules and regulations of both Federal and State medical transportation reimbursement regulations.

In 2010, the City of Memphis Division of Fire Services signed a contract with a private EMS billing service to process 80,000 transports annually. The contract also called for the private service to implement a complete field data-collection service including electronic Patient Care Reporting (ePCR) software set up for wireless communication from vehicles. The system was anticipated to save the city $175,000 annually in fees alone. Motivation for the change was a need to transition from a paper-based system to an electronic one, with the aim of improving the level of patient care, while increasing ambulance-reimbursement revenue.

Outsourcing won't always be the right answer, but it should be something EMS chiefs consider when looking for ways to save money.

Leasing

Leasing is a common financing strategy for many businesses with billions of dollars of equipment leased annually. Leasing allows an organization to acquire equipment and technology needed today, while spreading payments out affordably over time. This also allows an agency to reserve capital to finance daily operations or other expenses.

Leasing is much less used in the fire service because most fire apparatus is custom built. However, many businesses lease ambulances and fleet purchasers can supply sport utility vehicles (SUVs) to EMS agencies for use as command vehicles or paramedic intercept cars.

Tax-exempt, lease-purchase financing is a variation on the traditional lease. It offers several benefits:

- requires no down payment;
- tax-exempt interest rate;
- allows an agency to pay for assets over time;
- builds equity in the leased equipment with the first payment;
- at the end of the lease term, the department owns the equipment;
- flexible payment terms; and
- eliminates the paperwork and voter-approval required under debt financing.

In a lease-purchase agreement, however, an agency can end up paying far more for the vehicle than if they had purchased the vehicle outright. The terms of leasing need to be compared to a traditional purchase or the use of bonds, COP, or other debt-financing mechanisms.

Nontraditional Fundraising Ideas

Fundraisers are still a major part of alternative funding for many rural and volunteer fire and EMS departments. Ideas for fundraisers are limited only by the imagination. These fundraising events also provide the opportunity to present information on public life safety, health, and wellness. Don't try to compete with the local schools by selling candy or cookies. Be creative. Here are just a few ideas.

50/50 Raffle

The perfect fundraiser to hold at any event where there will already be a crowd of people, such as a sporting event or county fair. Tickets can be sold for $1 to $5. When a winner is drawn, that person receives half of the total dollar amount collected; the department keeps the other half.

5k, 10k, and Fun Run Event

People of all ages and athletic ability can take part in a fun run/walk event. It's a great way to raise money while promoting health and fitness. In addition to revenue raised from registration fees, consider combining this event with others such as a 50/50 raffle, barbecue lunch, health snack sales, t-shirt sales, and others.

Barbecue Fundraiser

A traditionally popular fundraiser, but requires a lot of volunteers and a few good cooks. Sponsors can help provide food, grills and fuel, tables and seating, and tents, along with door prizes and children's activities.

Basketball Tournament

A great way to raise money while promoting fitness! Three-on-three tournaments are a great way to raise money. This fundraiser is relatively inexpensive to setup and money is raised through entry fees. Earn extra money setting up a concession stand, selling photos and shirts, and selling ad space. Work with the local health club, YMCA, or school district for suitable basketball courts. Consider other sporting event tournaments like bowling, golf, or flag football.

Bingo

It offers players the chance to win prizes, but it's also a social outing. Bingo is a game popular among seniors. Consider setting up health-screening booths at the bingo event and offer blood pressure checks, glucose readings, vaccinations, and health screenings. The agency may not raise money for these services, but it's a great way to raise goodwill along with extra cash.

Bottled Water

Bottled water with your custom label can be sold year-round and is a walking billboard for your agency. As Americans become more health conscious, candy, cookies, pizza, and other foods are tougher to sell (especially for a public health agency like EMS).

First-Aid Kits

Run a drive offering small first-aid kits for home or car. It is a good idea to take orders before making a bulk purchase. Ask vendors if they can personalize the kit, adding your department's logo and message. What better way for an EMS provider to fundraise?

Fundraising Bricks
These are a great way to raise funds for new construction projects. Engraved bricks can be used for fundraising and recognize the donor at the same time. It is not uncommon for bricks to sell for several hundred dollars, even a thousand dollars or more. Auction off bricks in prime locations or group bricks for larger donors.

Car Wash
A staple of many volunteer fire departments; a car wash is quick and easy to setup.

Catalog
Catalogs have become one of the biggest moneymakers for fundraising. Make sure to select the right catalog depending on the season. It should have good merchandise at reasonable prices.

Coffee Fundraiser
Hosting a coffee fundraiser is a trendy idea. The agency can setup a "coffeehouse" at the station for people to visit and learn more about the organization. Bags of ground or whole-bean coffee can be sold to raise even more funds or delivered for those who can't attend in-person. Most people don't question $8–$10 per pound.

Cookbooks
Firefighters are also known (fairly or not) to be good cooks. Have the department put together a few favorite recopies from its members, take some photos, and put together a cookbook to sell.

Coupon Books
Coupon books are very popular in North America and are another big-seller. The coupons have to be a good value for the money to get people to buy them.

Custom T-shirts and Other Items
A lot of people like to wear or use items from public safety organizations. These can be popular fundraising items. However, do not sell any item that might look like a uniform.

Direct-Mail Campaign
This is raising money the old fashioned way: asking for it. Direct-mail fundraising can be effective, but it does have high overhead costs including the cost of printing, materials, and postage.

Email Campaign
Email campaigns are much cheaper and easier to run than direct mail. Consider putting together a monthly email newsletter with a permanent link to a donation website.

Healthier Fundraising Snacks
These days, people want healthier choices. Healthy snacks offer an opportunity for the fire/EMS agency to raise funds as well as awareness of healthy eating habits.

Kid Safety IDs
Kid safety ID fundraisers not only raise funds, they also serve a great public safety purpose. Child safety fundraisers work best when they are geared toward parents with preschoolers and elementary school-aged children. Kits retail from $8 to $25 with the organization netting about 40 percent.

Seasonal Fundraising Events

There are a number of fundraising ideas for each season. Consider hosting a haunted house for Halloween. Hold a turkey raffle for Thanksgiving. Christmas ideas include wreath sales, calendars, and holiday cards.

Vending Machines

These machines have a very high-profit margin. Both new and used vending machines are available online. It's a great way to raise money year round.

Wait Tables

Ask your local restaurant to allow members to wait tables and collect tips as a fundraiser. Some restaurants may even provide a matching donation or donate a portion of the night's sales to the fire/EMS agency.

Website and Internet Fundraising

Most departments have a webpage. Why not give visitors the opportunity to financially support the organization? Have the Webmaster add a page for credit card donations that deposit directly to the agency's bank account.

Things to Consider

Volunteer organizations may have an advantage in being able to raise money with fewer restrictions than career departments. Donations to the volunteers are usually tax deductible. Some career departments establish foundations to receive tax-deductible donations for special purposes. However, volunteer organizations need to be aware that they are still regarded as a public service entity and may be held accountable for the methods they use to raise funds and how the money is spent. Volunteers must abide by State and local ordinances, which may vary considerably across the country. For example, bingo may be legal in one district but illegal in another; and jurisdictions may have different rules about financial disclosure.

Fire/EMS departments should carefully consider the costs versus benefits of taking on new fundraising projects. Year-round fundraising activities such as bottled water sales or vending machines are time-consuming for members, whereas an annual fund drive may be easier to take on as a project.

The clarity and content of the fund-raising message to the public must also be given careful thought. Will the money be earmarked for a specific project, such as the purchase of an automated external defibrillator (AED) or fire truck? Or will the money go towards operational expenses? It is typically easier to raise money for a specific project than for general operational costs.

Fire/EMS departments should tailor their fundraising activities to their community characteristics and preferences. Would a small community respond best to door-to-door efforts? Would a larger, more affluent community respond better to a direct-mail campaign? Are the local citizens likely to attend a bingo event or would they respond better to a barbecue?

Departments should consider getting the local media involved in the fundraising effort. The press can help agencies distribute flyers, produce newspaper articles highlighting the event, and create advertisements from print, radio, and cable television. To get the word out, agencies may want to consider presentations at local civic and religious organizations to solicit on behalf of the department. If the agency has tax-exempt status, consider reaching out just before the end of the year to allow a donor to make a tax-deductible donation before the tax year concludes. Be careful not to constantly bombard the community or they could get the impression that the department is only in business to raise money.

There are many community resources that fire/EMS departments can use to assist them in their fundraising efforts. For example, local businesses may be willing to advertise promotional messages for fundraising campaigns. Schools may be willing to let departments send flyers home with children. Local celebrities may be willing to endorse the efforts as well.

Chapter 5: State Funding for Emergency Medical Services

Much of the funds dispersed to emergency medical services (EMS) organizations by States come from Federal passthrough grants such as the Community Development Block Grants or the State Homeland Security Grant Program. These grants are managed within various State agencies and transferred to local or tribal governments and non-fire-based EMS organizations. Good sources of information on available grants can be found in State Departments of Health, Departments of Transportation (DOTs), Departments of Natural Resources, Insurance Departments, State Fire Marshal's Office, and Department of Emergency Management and/or Homeland Security, among others. States may also have their own intrastate mechanisms to provide funding for EMS.

This section provides a list of alternative revenue sources available from State agencies, both passthrough and State funding programs. Although every effort was made to compile a comprehensive listing of programs, newer programs may not be presented here. Regardless, this list should provide agencies looking for alternative funding with a good starting point for information about existing programs. Further, by seeing the options available in other States, fire and EMS agencies may be able to encourage the development of new programs in their own State.

State Sources of Revenue

In addition to administering the dispersal of Federal funds, many States provide direct funding through grants, loans, or technical support for local EMS. These programs are financed through dedicated funding streams such as insurance surcharges or driving-related fees and fines. State governments may supply support services such as loans, surplus equipment, special-purpose grants, matching grants, technical assistance, and training.

Low-Interest Loans

Many States may make funds available through low- or zero-interest loans, with favorable terms of 20 years or more, for capital improvements and purchases. Such programs are often setup in a revolving loan fund so that money paid back to the lender can be loaned out again to other fire/EMS agencies. One example, the Illinois Office of the State Fire Marshal and the Illinois Finance Authority jointly administer a program to provide zero-interest, revolving loan program for the purchase of ambulances by not-for-profit EMS or local government agency that provides EMS. Money is lent on a 10-year repayment term.

Surplus Vehicles and Equipment

Secondhand fire apparatus, ambulances, and equipment suitable for rescue squads are available through the State at little or no cost, especially for rural and volunteer departments. The challenge is finding them. Some may be available through the State forest service, others from the National Guard. Many States have offices specifically responsible for the disposition of State assets. The State Procurement Office may be able to assist agencies in locating used State equipment.

States may also have contracts with local dealerships guaranteeing lowest price. Fire departments and EMS agencies may be able to purchase vehicles at State contract prices.

Special-Purpose Grants

From time to time, some States earmark funds for a particular program area of interest. Additionally, excess funds are sometimes available from the State legislature at the end of the fiscal year. These funds may be made available in the form of grants to various organizations in the State. Contact the local State representative or State senator to inquire about the availability of such funds and the process involved to apply for them.

Matching Grants

Matching grants are a common source of funds for local agencies at the State level. While either the Federal or State government may fund these programs, the dispersal of funds is generally left up to the State agency tasked with managing the grant. These grants require the department to match a portion of the grant. Depending on the grant guidance, this can be cash or in-kind contributions. Rural and volunteer organizations may receive a discounted rate on matching grants.

For example, in Oklahoma, the State legislature appropriates funds each year for grants and cost-share programs administered by the Oklahoma Forestry Services, a Division of the Department of Agriculture, Food, and Forestry. These programs provide financial aid and matching grants for rural fire departments to construct facilities or purchase equipment. The Rural Fire 80/20 Reimbursement Grant reimburses rural fire departments for up to 80 percent of the cost of equipment purchases or fire station projects.

The Florida Department of Health, Bureau of Emergency Medical Services, offers two types of matching grants. The General Matching Grant is available to private, urban and rural organizations involved in EMS where the State pays 75 percent of the approved budget and the grantee pays 25 percent. The Rural Matching Grants are available to public and private organizations involved in EMS which are based in and provide services for counties with a population of 100,000 or fewer people and a population density of less than 100 per-square-mile. For this grant, the State pays 90 percent of the approved budget and the grantee only 10 percent.

Technical Assistance

The State Fire Marshal's Office and Bureau of Emergency Medical Services may be sources of programmatic support, technical assistance, and reporting, and a centralized database information collection to assist fire departments and EMS agencies in accomplishing their public health and safety missions. Often, State EMS offices provide services such as trauma and critical care informatics, regional coordination, regulations and compliance, assistance in writing quality assurance plans, and guidance documents for treatment protocols, EMS curricula, etc. The State Fire Marshal is a source of technical support to local fire departments on fire protection engineering questions, fire code interpretations, and opinions on the application of fire safety laws and regulations. The State Fire Marshal's Office may also help collect and analyze data to determine the causes of fire and burns and assist with the development of appropriate prevention plans. A few State Fire Marshal Offices provide fire investigative services as well.

Subsidized Training

Many States subsidize the cost of fire and EMS training, particularly for rural and volunteer departments. The Ohio Department of Commerce Division of State Fire Marshal provides grant funding for fire department reimbursement of State-certified fire classes conducted at the Ohio Fire Academy. Training grants and the availability of certified fire training varies significantly from State to State. Chief fire officers are advised to contact their State Fire Marshal's Office for details and availability.

Across the country, State EMS offices oversee the training and certification of emergency medical technicians (EMTs) and paramedics. Some States provide grants or scholarships to qualified individuals, especially rural or volunteer EMTs. The Texas Department of Rural Affairs maintains a Rural Emergency Medical Services/Trauma Enhancement Grant. The purpose is to develop and improve rural EMS response, delivery, and service capacity. Grants are awarded for EMT training, continuing education courses, and certification of instructors, among other uses.

In addition, the Texas DOT, Traffic Safety Division, provides a Rural/Frontier EMS Education Fund in conjunction with the Texas Engineering Extension Service (TEEX). The fund covers tuition for initial, refresher, and recertification of EMT, EMT-I, and paramedics. TEEX also hosts continuing education courses such as advanced cardiac life support (ACLS) and emergency medical dispatch (EMD).

State-Managed Federal Grants

Opportunity Title: Community Development Block Grant (CDBG)

Sponsor: U.S. Department of Housing and Urban Development (HUD).

Purpose: CDBG is a flexible program that provides communities with resources to address a wide range of unique community development needs. The CDBG program provides annual grants on a formula basis to States and local governments.

Eligible Activities: Seventy percent of CDBG funds must be used for activities that benefit low- and moderate-income persons. In addition, each activity must meet one of the following national objectives: benefit low- to moderate-income persons, prevention or elimination of slum and blight, or address community development needs having a particular urgency because existing conditions pose a serious and immediate threat to the health or welfare of the community for which other funding is not available.

Eligibility: The CDBG entitlement program allocates annual grants to central cities of the Metropolitan Statistical Areas, cities of at least 50,000, and qualified urban counties over 200,000. States distribute CDBG funds to nonentitlement localities.

HUD provides flexible grants to help cities, counties, and States recover from Presidentially-declared disasters, especially in low-income areas, subject to availability of supplemental appropriations.

Texas, Arizona, California, and New Mexico set aside 10 percent of their State CDBG funds for improving living conditions for colonial residents. The Insular Areas CDBG program provides grants to four designated insular areas: American Samoa, Guam, Northern Mariana Islands, and the Virgin Islands.

More Information: www.hud.gov/offices/cpd/communitydevelopment/programs

Opportunity Title: Rural Fire Assistance (RFA) Grant Program

Sponsor: U.S. Department of the Interior (DOI).

Purpose: The RFA Grant Program is designed to increase firefighter safety and enhance the fire protection capabilities of rural fire departments. Emphasis is placed on departments that protect communities of less than 10,000 and play a substantial role in the protection of Federal lands managed by the U.S. DOI.

Eligible Activities: Funding assistance is available for wildland fire personal protective equipment (PPE), wildland fire-hand tools, new generation fire shelters and training, Project 25-compliant radios, basic wildland fire training, and apparatus purchased from local DOI units through "negotiated fixed price sale."

Eligibility:

- Statewide agreement with State Forester or cooperative fire agreement with an Interior Bureau;
- fire department with a service area of 10,000 or less;

- a fire department with at least one station that serves a community of more than 10,000 that also encompasses a rural zone or community; and/or
- a fire department that serves a community of 10,000 or more that also provides fire protection services through contract or agreement to an adjoining rural community.

More Information: www.nifc.gov/rfa/

Notes: The maximum allowable Federal contribution for each cooperating rural fire department is $20,000. The Federal/Co-operator cost-share ration cannot exceed 90/10; the smaller portion being the cooperator's responsibility. Cooperator contributions may be in the form of in-kind service.

Opportunity Title: Volunteer Fire Assistance (VFA) Grant

Sponsor: U.S. Department of Agriculture (USDA), U.S. Forest Service (FS).

Purpose: Provides financial assistance to train, organize, and equip fire departments in rural areas and rural communities to prevent and suppress fires.

Eligible Activities: Training and expenses related to training activities, costs associated with organizing a fire district, risk assessment and hazard mitigation planning, fire tools, PPE, equipment renovation, self-contained breathing apparatus (SCBAs), hoses and nozzles, purchase or lease of fire apparatus, etc. Activities that cannot be funded through the grant include EMT or paramedic training and equipment.

Eligibility: Rural community with a 10,000 or less population.

More Information: www.fs.fed.us/fire/partners/vfa/

Notes: A requirement of the VFA Grant Program is a 50/50 match for the Federal funds on a reimbursable basis. Departments must spend their matching funds or provide in-kind services prior to receiving the grant funds. The maximum grant amount for each project is $20,000.

Opportunity Title: Western States Wildland Urban Interface (WUI) Grant Program

Sponsor: USDA, FS.

Purpose: The focus of this funding is mitigating risk in WUI areas. Funding is available to western States and awarded through a competitive process with emphasis on hazard-fuel reduction, information and education, and community and homeowner action.

Eligible Activities: Reduce hazardous fuels/restore fire-adapted ecosystems (defensible space around homes and structures, prescribed fire, thinning), improve prevention/education in the WUI (Firewise programs, Project Learning Tree, pamphlets, and promotional materials), and creation of Community Wildfire Protection Plans.

Notes: The grant amount must be matched 50/50 by the recipient using a non-Federal source. Title III funds under the Secure Rural Schools and Community Self-Determination Act of 2000 (www.fs.usda.gov/) are not considered Federal dollars and may be used as a match.

Specific State Grants and Aid Programs

In all cases, legislative appropriations to grant programs administered by State agencies are not guaranteed nor should they be solely relied upon. Applicants are strongly encouraged to consider the long-term sustainability of their programs and seek alternative sources of support and funding. The grant programs presented in this section may be cancelled or unfunded at any time. Consult the administrative contact provided for current information.

Alabama

Opportunity Title: Wildland Fire Prevention Grant

Sponsor: Alabama Forestry Commission.

Purpose: To assist the Alabama volunteer fire service in reducing the number of wildfires and resulting structural loses, while increasing the visibility and importance of volunteer fire departments in their communities.

Eligible Activities:

- public relations and education;
- leverage private funds and corporate sponsorships for wildland fire prevention activities;
- community wildfire prevention planning; and/or
- coordinate wildland fire prevention programs with the local Alabama Forestry Commission.

Eligibility: Volunteer fire departments organized as part of a municipality or nonprofit organization; county fire associations organized as a nonprofit organization; and District fire associations organized as a nonprofit.

Website: www.aavfd.org/admin.html

Contact: Alabama Association of Volunteer Fire Departments
550 Adams Avenue, Suite 345
Montgomery, AL 36104
(334) 262-2833
Fax: (334) 262-2834

Alaska

Opportunity Title 1: Alaska Code Blue Project

Sponsor: Health and Social Services, Division of Public Health, Emergency Programs.

Purpose: The Code Blue Project was started in 1999 in an effort to identify, prioritize, and seek funding for essential equipment for rural EMS in Alaska. The program is coordinated by a Steering Committee comprised of a representative from several State agencies. The program has resulted in the purchase of over 25 new ambulances, rescue vehicles, patient care equipment, training equipment, and more.

Eligible Activities: Items are reviewed at the regional and State levels prior to inclusion in the Code Blue Database. General categories of equipment within the database include

- patient care;
- training;
- emergency vehicles;
- radio/communications equipment;
- ambulances;
- emergency equipment; and
- communications infrastructure.

Eligibility: Rural EMS providers in Alaska.

Website: www.chems.alaska.gov/EMS/Programs/code_blue.htm

Contact: EMS Programs
410 Willoughby Avenue, Room 103
P.O. Box 110616
Juneau, AK 99811-0616
Main: (907) 465-1733
Certification: (907) 465-6736
EMS: (907) 465-4101

Opportunity Title 2: Alaska Highway Safety Office (AHSO) Grant

Sponsor: AHSO.

Purpose: To reduce the number of injuries and deaths resulting from traffic crashes on Alaska's public roadways. The AHSO prefers innovative projects that provide solutions to identifiable problems, implements proven strategies, show a commitment on behalf of the applicant to sustain and contribute to success, and have measurable outcomes.

Eligible Activities:

- public education campaigns;
- equipment and materials;
- traffic record systems; and
- training and travel directly related to the project.

Eligibility: Includes EMS.

Website: www.dot.state.ak.us/stwdplng/hwysafety/grantinfo.shtml

Contact: Alaska Highway Safety Office
P.O. Box 112500
3132 Channel Drive
Juneau, AK 99811-2500
Fax: (907) 465-4030

Notes: Projects that include matching funding receive special considerations. Grants are awarded on a "cost-reimbursement" basis. Grantees are required to submit reimbursement vouchers on a monthly basis.

Arizona

Opportunity Title: Arizona Governor's Office of Highway Safety (GOHS) Grant

Sponsor: Arizona GOHS.

Purpose: GOHS funds highway safety projects using Federal 402 State and Community Highway Safety Grants, 410 Alcohol Incentive, 408 Information Systems Improvement, and Motorcycle Safety funds. Additionally, GOHS receives funding through the U.S. Department of Justice (DOJ) for Underage Drinking Enforcement.

Eligible Activities: The GOHS grant supports the Arizona Highway Safety Plan in specifically-targeted areas including impaired driving, speed and police traffic services, and occupant protection. Proposals are also accepted for Emergency Medical Services, Motorcycle Safety programs, and Pedestrian and Bicycle Safety programs.

Eligibility: For Fire and EMS Departments: Per National Highway Traffic Safety Administration (NHTSA), as long as the proposal is traffic-safety related, it will be evaluated and reviewed for perspective funding by both GOHS and NHTSA. Unfortunately, there is no established list of allowable projects and each proposal will be evaluated on a case-by-case basis.

Website: www.azgohs.gov/grants-opportunities

Contact: Arizona Office of Highway Traffic Safety (602) 255-3216.

Arkansas

Opportunity Title 1: Act 833 Fire Grant Program

Sponsor: Arkansas Department of Emergency Management.

Purpose: Act 833 of 1991 is to provide funding for qualified fire departments and is not intended for daily or operation funding. It must be directly related to firefighting capabilities and can only be used for training, equipment procurement, and capital expenditures. Funds are distributed to each county based on population.

Eligible Activities: Approved fire training courses through the Arkansas Fire Training Academy, National Fire Academy (NFA), Arkansas Department of Emergency Management, and/or fire training centers, colleges or universities, and approved by the Arkansas Fire Protection Services Board. Procurement of new or modernization of current firefighting equipment; and capital expenditures and/or security pledges (10 percent or less) to be used in financing of firefighting equipment, initial capital construction, or improvement of fire departments including purchase of property, construction of stations, and firefighting apparatus.

Eligibility: Fire departments.

Website: www.adem.arkansas.gov

Contact: Arkansas Fire Protection Services Board C/O
Office of Fire Protection Services
Arkansas Department of Emergency Management
Building #9501–Camp Joseph T Robinson
North Little Rock, AR 72199-9600
(501) 683-6781

Notes: Act 833 funds are subject to audits; therefore, they must be used in accordance with Act 833 rules and regulations. Beginning in 2008, random, onsite audits will be conducted on fire departments which receive or have received Act 833 funds.

Opportunity Title 2: Rural Community Development/Rural Fire Protection Grant Program

Sponsor: Arkansas Department of Rural Services.

Purpose: Applicants from incorporated towns of less than 3,000 in population and unincorporated rural areas are eligible for up to $15,000 in matching funds under this program. The match ratio on the program is 50/50. Communities wishing to apply for projects under the program must provide one-half of the cost of the project as match. Match may be comprised of in-kind labor, in-kind materials, or cash, and must be available at the time of application. Applications for this program are accepted August through March of each year.

Eligible Activities: The Arkansas Rural Services Block Grant Program (RSBGP) funds new construction or renovation of community centers, fire stations, or multipurpose buildings, and the purchase of fire trucks (pumper, tanker, brush, or certain service trucks). This grant is made possible through a partnership with the Arkansas Economic Development Commission.

Eligibility: Incorporated and unincorporated cities and towns in rural Arkansas with less than 3,000 population; population must be 51 percent low- to moderate-income (LMI); and is not within the outer boundary of any city or town having a population of 20,000 or more.

Website: www.arkansas.gov/drs/index.html

Contact: Department of Rural Services
Attn. RSBGP Coordinator
101 East Capitol, Suite 202
Little Rock, AR 72201
(888) 787-2527

Notes: The RSBGP is a nine-to-one matching program. This requires that the applicant must be able to match its grant award with cash, in-kind labor, in-kind materials, or land at a rate of one dollar to every nine grant dollars (10 percent). Land used for matching purposes must be owned by the city or county applying.

Opportunity Title 3: Community Fire Prevention Grant

Sponsor: Arkansas Department of Emergency Management.

Purpose: The purpose of this grant program is to provide needed financial support for fire departments, community or faith-based organizations to start, enhance or expand their fire and life safety programming.

Eligible Activities: Fire prevention-related activities. Activities must endeavor to reduce the loss of life or property due to fire. Organizations must demonstrate a plan to implement a community-wide fire and life safety program/campaign and the program/campaign must be aimed at the general public or targeted groups at risk in the organization's area.

Eligibility: Only Arkansas-based, legally recognized organizations representing the local, regional, or Statewide interests of the fire service shall be eligible as well as community or faith-based organizations with a letter of collaboration from appropriate fire service entity. Must be National Fire Incident Reporting System (NFIRS) reporting community.

Website: www.adem.arkansas.gov

Contact: Arkansas Fire Prevention Commission
Arkansas State Fire Marshal
#1 State Police Plaza
Little Rock, AR 72010
(501) 618-8624

Notes: Use of grant money toward fire suppression or recreational activities is expressly prohibited.

California

Opportunity Title 1: Hazard Mitigation Grant Program (HMGP)

Sponsor: California Emergency Management Agency (CalEMA).

Purpose: The HMGP program provides grants to State and local governments to implement long-term hazard mitigation measures after a major disaster declaration. Federal funding is provided under the Robert T. Stafford Emergency Assistance and Disaster Relief Act through Federal Emergency Management Agency (FEMA) and the CalEMA.

Eligible Activities: Funding is available for hazard-mitigation projects for severe winter storms, flooding, and debris and mud flows, or the development of all-hazard mitigation plans.

Eligibility: State and local governments, special districts, public colleges and universities, and certain private nonprofit organizations.

Website: www.calema.ca.gov

Contact: California Emergency Management Agency
Disaster Assistance Program Branch
Hazard Mitigation Section–Attention 1557/1585
3650 Schriever Avenue
Mather, CA 95655
(916) 845-8150

Notes: Funds are provided on a 75/25 cost-share basis; 75 percent Federal and 25 percent non-Federal. Matching funds must be identified and secured in the application when submitted. The Federal share may not exceed $3 million for projects, $150,000 for multijurisdictional plans and updates, and $75,000 for single-jurisdiction plans and updates.

Opportunity Title 2: California Traffic Safety Grant

Sponsor: California Office of Traffic Safety.

Purpose: The California Office of Traffic Safety administers grants to fund projects and programs to reduce traffic deaths, injuries, and economic losses.

Eligible Activities: EMS expenses including ambulance and equipment purchases up to 25 percent of the total cost. Grants can be used to mitigate traffic safety program deficiencies, expand on going activities, or develop new programs.

Eligibility: Government agencies.

Website: www.ots.ca.gov/grants/default.asp

Contact: California Office of Traffic Safety
2208 Kausen Drive, Suite 300
Elk Grove, CA 95758
(916) 509-3030

Notes: Grant funding cannot replace existing program expenditures, nor can traffic safety grants be used for program maintenance, research, rehabilitation, or construction.

Opportunity Title 3: Fire Safe California Grants Clearinghouse

Sponsor: Fire Safe California.

Purpose: Fire Safe California provides a one-stop shop that simplifies the process of finding and applying for grants to improve California's community wildfire preparedness.

Eligible Activities: Funds are available to assist with hazardous fuels treatments, community wildfire protection planning, and education addressing wildfire safety and hazardous risk-reduction within the WUI.

Eligibility: Organizations working in California and representing their communities including State, local, county, interstate, intermunicipal, special districts, independent school districts, State-controlled institutions of higher learning, private colleges and universities, Native American communities, and eligible for-profit and nonprofit organizations (must have a Federal EIN and be registered as a business).

Website: www.grants.firesafecouncil.org

Contact: California Fire Safe Council
502 West Route 66, Suite 17
Glendora, CA 91740
1-800-FS CALIF (372-2543)

Colorado

Opportunity Title 1: Emergency Medical and Trauma Services Provider Grant Program

Sponsor: Colorado Department of Public Health and Environment.

Purpose: The EMS and Trauma Provider Grant Program assists private and public organizations in improving and expanding the emergency medical and trauma system in Colorado.

Eligible Activities: Funding is available for the purchase of ambulances and equipment, communications, data collection, injury-prevention programs, and personnel services and recruitment efforts.

Eligibility: Applicants must have their purpose of provision of emergency medical and trauma services in Colorado to be eligible.

Website: www.cdphe.state.co.us/em/grants

Contact: Colorado Department of Public Health & Environment
HFEMSD-EMTS-C1
4300 Cherry Creek Drive South
Denver, CO 80246-1530
(303) 692-2980
Fax: (303) 691-7720

Notes: The grant program does not fund land acquisition, capital construction, fire apparatus, uniforms, disposables, and infrastructure.

Opportunity Title 2: Colorado Rural Health Care Grant Program

Sponsor: Colorado Rural Health Center.

Purpose: The general purpose of this program is to improve the provision of health-care services in rural Colorado. The funds are intended to support outpatient primary care/outpatient care is considered to be the basic, entry-level care provided by clinicians.

Eligible Activities: Equipment, construction, physical plant improvements, vehicles and vehicle upgrades, information technology, and staff training or education.

Eligibility: Only applicants from rural Colorado are eligible to apply. Rural applicants have the option to contract for urban-based services or resources to meet the objectives of their grant proposals.

Website: www.coruralhealth.org/programs/crhcgp

Contact: Colorado Rural Health Center
3033 South Parker Road, Suite 606
Aurora, CO 80014
(800) 851-6782

Notes: The maximum award amount is $50,000 per applicant.

Opportunity Title 3: Colorado Resource for Emergency and Trauma Education (CREATE)

Sponsor: Colorado Rural Health Center.

Purpose: CREATE assists private and public organizations in improving and expanding the emergency medical and trauma system in Colorado.

Eligible Activities: Education.

Eligibility: Applicants must have as their purpose the provision of emergency medical and trauma services in Colorado to be eligible.

Website: www.coruralhealth.org/programs/create/index.htm

Contact: Colorado Rural Health Center
3033 South Parker Road, Suite 606
Aurora, CO 80014
(800) 851-6782

Opportunity Title 4: EMS and Trauma Emergency Grants

Sponsor: Colorado Department of Public Health and Environment.

Purpose: Colorado's emergency grant program assists private and public EMS and trauma providers who experience emergencies that seriously jeopardize the level of EMS available within their service area.

Eligible Activities: Applicants must have as their purpose the provision of emergency medical and trauma services in the State of Colorado to be eligible.

Eligibility: In addition to meeting the standard eligibility requirement, the emergency fund applicant must be able to demonstrate that the request for funds is a true emergency and, if funds are not received, will degrade the quality of, or eliminate access to, local EMS and trauma care.

Website: www.cdphe.state.co.us/em/grants/emergencyinfo.html

Contact: Colorado Department of Public Health & Environment
HFEMS-A2
4300 Cherry Creek Drive South
Denver, CO 80246-1530
(303) 692-2980
Fax: (303) 691-7720

Connecticut

Opportunity Title: Connecticut Fair Access to Insurance Requirements (FAIR) Plan Committee Anti-Arson Grant

Sponsor: Connecticut FAIR Plan Committee.

Purpose: The FAIR Plan Anti-Arson Committee has established a grant fund from which Connecticut State and local governmental agencies can request equipment relating to the prevention, investigation, or prosecution of arson fires.

Eligible Activities: Small items of equipment or material that would aid in fire investigation or the prosecution of arson. There must be sufficient need to justify the expense and qualified personnel available to operate and maintain the equipment.

Eligibility: Requests must come from official State or local fire marshal offices, fire departments, police departments, or the State's Attorney's offices.

Website: www.ctfairplan.com

Contact: Connecticut FAIR Plan
Anti-Arson Committee
P.O. Box 280200
East Hartford, CT 06128-0200
(860) 528-9546

Notes: The maximum grant will not exceed $500 in any 12-month period.

Florida

Opportunity Title 1: Florida EMS County Grant Program

Sponsor: Florida Department of Health.

Purpose: The Department of Health, Bureau of Emergency Medical Services is authorized to dispense grants. Forty-five percent of these funds are made available to the 67 boards of county commissioners (BCCs) to improve and expand prehospital EMS systems in their county.

Eligible Activities: EMS vehicles and equipment purchases.

Eligibility: EMS county grants are awarded to BCCs.

Website: www.fl-ems.com/Grants/Grants.html

Contact: County Grant
Emergency Medical Services
4052 Bald Cypress Way, Bin C18
Tallahassee, FL 32399-1738
(850) 245-4440

Opportunity Title 2: Florida Highway Safety Grant Program

Sponsor: Florida DOT.

Purpose: The Traffic Safety Grant Program provides seed funding to solve traffic safety problems from educational and enforcement approaches. Topics addressed include alcohol, seatbelts and child-safety seats, police traffic services, speed and aggressive driving, traffic records, motorcycle, roadway, community traffic, and pedestrian and bicycle safety.

Eligible Activities: Funding for State and local programs. Also, some funding of the emphasis areas has resource centers that provide technical assistance, publications, and safety materials.

Eligibility: State agencies, universities, units of local government, and nonprofit organizations.

Website: www.dot.state.fl.us/default.shtm

Contact: Florida Department of Transportation
1211 Governor's Square Boulevard
Magnolia Center 2, Suite 300
Tallahassee, FL 32301
(850) 245-1504

Notes: Since these grants are intended to be seed funding, local agencies are expected to pick up some of the program costs, especially for any personnel, in the second and third years of funding.

Georgia

Opportunity Title: Georgia Local Law Enforcement and Fire Services State (LLEFS) Grant

Sponsor: Georgia Criminal Justice Coordinating Council.

Eligible Activities: General purpose fire and police grants.

Eligibility: By the nature of this program, only local government agencies are eligible to apply.

Website: www.cjcc.ga/gov

Contact: (404) 657-1963

Hawaii

Opportunity Title: Hawaii Highway Safety Grant Program

Sponsor: Hawaii DOT.

Purpose: The grant supports State highway safety programs, designed to reduce traffic crashes and resulting deaths, injuries, and property damage.

Eligible Activities: EMS—To ensure appropriate treatment through a coordinated system of emergency medical care for persons injured in highway crashes.

Eligibility: The grant funds are for traffic safety purposes only.

Website: www.hawaii.gov/dot

Contact: (808) 587-6301

Notes: The State of Hawaii has developed priority levels for funding based on data from highway collisions in the State. Priorities reflect the number of injuries and deaths each year. As a result, Hawaii has deemed equipment purchases by EMS to be a "lower priority" for funding.

Idaho

Opportunity Title 1: Idaho Highway Safety Grant Program

Sponsor: Idaho Transportation Department.

Purpose: The goal of the program is to reduce deaths and injuries from motor-vehicle crashes through funding programs and activities that promote safe travel on Idaho's transportation systems, and through collecting, maintaining, and disseminating reliable crash statistics. Projects that are considered for funding must address the emphasis areas identified in Idaho's Strategic Highway Safety Plan.

Eligible Activities: Crash Response (EMS)—The goal of this program area is to enhance appropriate, timely, and safe response to crashes to reduce the time that it takes first responders to remove injured crash victims from the crash site and transport them to advanced medical treatment. Funding priorities for this area are for the purchase of extrication equipment and/or educational opportunities.

Eligibility: State and local governmental units and nonprofit organizations.

Website: www.itd.idaho.gov/ohs/programs.htm

Contact: Idaho Transportation Department
Office of Highway Safety
P.O. Box 7129
Boise, ID 83707-1129
(208) 334-8100

Opportunity Title 2: Communities at Risk

Sponsor: Idaho Department of Lands.

Purpose: Provides financial assistance to local jurisdictions in Idaho for efforts that support fire prevention activities.

Eligible Activities: Funds may be used for planning efforts (including the use of Geographic Information System (GIS) software and support), the hiring of countywide WUI coordinators, and education efforts such as FIREWISE. Funds may also be used to reduce hazardous fuels accumulations on non-Federal lands, however, use of funds for this purpose may require environmental clearance.

Eligibility: County Wildland Fire Interagency Groups, county governments, communities, nonprofit entities.

Website: www.idahofireplan.org/grants

Contact: Idaho Department of Lands
(208) 373-3854

Opportunity Title 3: Community Fire Protection (formerly "Steven's Funds") and Bureau of Land Management (BLM) Partnership Fund

Sponsor: Idaho Department of Lands.

Purpose: Provide funding for hazardous fuels treatments on private lands that are adjacent to National Forest (Community Fire Protection) and BLM (Partnership Fund) property.

Eligible Activities: Funds may only be used for hazardous fuels work and not for related activities.

Eligibility: County Wildland Interagency Groups or county governments.

Website: www.idahofireplan.org/grants

Illinois

Opportunity Title 1: Emergency Medical Services Assistance Fund

Sponsor: Illinois Department of Public Health.

Purpose: The Emergency Medical Systems Assistance fund offers grant opportunities each year for EMS providers to purchase equipment and pay for training to help increase the safety of all Illinois residents.

Eligible Activities: EMS equipment and training.

Eligibility: EMS providers.

Website: www.idph.state.il.us

Contact: Illinois Department of Public Health
535 West Jefferson Street
Springfield, IL 62716
(217) 785-2083

Opportunity Title 2: The Heartsaver Automated External Defibrillator (AED) Grant Program

Sponsor: Illinois Department of Public Health.

Purpose: The Heartsaver AED Fund is for the purpose of providing matching grants for the purchase of AEDs.

Eligible Activities: Purchase of AEDs.

Eligibility: Any school, Public Park District, Municipal Recreation Department, Conservation District, Forest Preserve District, College, or University in the State of Illinois.

Website: www.idph.state.il.us

Contact: Illinois Department of Public Health
535 West Jefferson Street
Springfield, IL 62716
(217) 785-2083

Opportunity Title 3: Illinois State Fire Marshal Grants

Sponsor: Office of the Illinois State Fire Marshal.

Purpose: The Office of the Illinois State Fire Marshal has several grant opportunities for fire departments and fire districts, including the New Fire District Grant, Arson Fines Grant, Training Reimbursement Grant, and the Small Equipment Grant.

Eligible Activities: Small equipment, arson equipment, and training.

Eligibility: Any Illinois fire department or fire district.

Website: www.state.il.us/osfm

Contact: Office of the Illinois State Fire Marshal
1035 Stevenson Drive
Springfield, IL 62703
(218) 785-4717

Notes: In order for a fire department or fire district to be eligible for any grant program from the Office of the State Fire Marshal, the department or district must be in compliance with the NFIRS Recording Program.

Indiana

Opportunity Title: Burn Care and Prevention Grants

Sponsor: St. Joseph Community Health Foundation.

Purpose: Accidents are the leading cause of burns. Over the years, donors provided these restricted funds to promote burn-prevention education and to help educate health-care professionals and first responders to emergency situations to treat severe burns with best care practices available. Grants typically range between $500 and $5,000.

Eligible Activities: Burn-prevention education, burn-treatment training.

Eligibility: Nonprofit agencies classified as 501(c)3, some governmental agencies involved in related service areas may be eligible for support.

Website: www.sjchf.org

Contact: The St. Joseph's Community Health Foundation
2826 South Calhoun Street
Fort Wayne, IN 46807
(260) 969-2001 x201

Iowa

Opportunity Title: Iowa West Foundation Grants

Sponsor: Iowa West Foundation.

Purpose: The foundation will consider grant applications covering a wide range of community needs with a focus on community development and beautification, economic development, education, and human and social needs.

Eligible Activities: Past grant recipients have received funding for building improvements, small equipment, protective equipment, and grass-fire apparatus.

Eligibility: Communities in southwest Iowa and eastern Nebraska, the primary focus being on Council Bluffs and Pottawattamie County with a secondary focus on surrounding counties.

Website: www.iowawestfoundation.org/index.cfm

Contact: Iowa West Foundation
25 Main Place, Suite 550
Council Bluffs, IA 51503
(712) 309-3003

Kansas

Opportunity Title: Fire Injury Prevention Program Grant

Sponsor: Kansas Department of Health and Environment (KDHE).

Purpose: The overall purpose of the Kansas Fire Injury Prevention Program is to make funds available for community programs to reduce injuries due to fires and burns.

Eligible Activities: Examples of KDHE activities include smoke-alarm installations, fire safety training and education, non-English-speaking public education programs, and participation in workshops, training, and other meetings.

Eligibility: Private or nonprofit community-based organizations, units of State government, and units of local government serving a population of 50,000 or less. Fire departments and other public safety agencies are particularly encouraged to apply.

Website: www.kdheks.gov/kfipp/index1.html

Contact: Kansas Department of Health and Environment
1000 Southwest Jackson, Suite 230
Topeka, KS 66612-1274
(785) 368-7290

Notes: Fire departments should accurately report fire incidents to the State Fire Marshal's Kansas Fire Incident Reporting System (KFIRS).

Kentucky

Opportunity Title: Search and Rescue Grant

Sponsor: Kentucky Division of Emergency Management.

Purpose: The purpose of this grant is to solve identified problems or concerns for Kentucky Rescue Squads. Priority for grant funding will be given to rescue squads that have no other means of funding, i.e., Federal grants, Kentucky Office of Homeland Security (KOHS) grants, etc., and to requests that fulfill unique needs within a community or region, and teams that do not charge for services.

Eligible Activities: Purchase of minimum equipment required in Chapter 39F and optional rescue equipment.

Eligibility: Kentucky Rescue Squads.

Website: www.kyem.ky.gov

Contact: Kentucky Division of Emergency Management
SAR Coordinator - 1025 Capital Center Drive, Suite 101
Frankfort, KY 40602
(502) 607-1601

Maine

Opportunity Title: Maine Highway Safety Grants

Sponsor: Maine Bureau of Highway Safety.

Purpose: The Highway Safety Grants are used to reduce deaths, injuries, and property damage resulting from traffic crashes. The Bureau of Highway Safety addresses those highway safety problems that relate to the behavioral factors of drivers, pedestrians, cyclists, and other highway users.

Eligible Activities: EMS—Within this category, funds are dedicated to support training, public information and education, equipment acquisition, staff support, and program administration. Funded projects can provide direct services and/or system improvements.

Eligibility: Fire departments and EMS agencies.

Website: www.maine.gov/dps/bhs/index.shtml

Contact: Maine Bureau of Highway Safety
164 State House Station
Augusta, ME 04333-0164
(207) 626-3840

Maryland

Opportunity Title: Maryland Institute for Emergency Medical Services Systems (MIEMSS) Radio Grant Program

Sponsor: MIEMSS.

Purpose: Under the Radio Grant Program, MIEMSS reimburses a county (combination) and/or volunteer company in the county, up to a specified amount for the purchase, and installation where appropriate, of new narrowband mobile and portable radios that meet the MIEMSS specifications. There is a specific mobile radio and a specific portable radio that can be purchased under the Radio Grant Program which are compliant with Federal Communications Commission (FCC) requirements and meet the operational needs of the Maryland EMS communications system. Under the Radio Grant Program, MIEMSS reimburses the grantee up to $1,300 for the purchase of a Motorola Model CDM 1250 mobile radio including, where appropriate, installation costs, and MIEMSS reimburses the grantee up $1,300 for the purchase of the Kenwood TK 390 portable radio, mobile charger and, where appropriate, installation.

Eligible Activities: Purchase of radios and broadband communications equipment.

Website: www.miemss.org/home/MIEMSSRadioGrantInformation/tabid/186/Default.aspx

Contact: 653 West Pratt Street
Baltimore, MD 21201
Phone: (800) 762-7157
Email: info@miemss.org

Notes: It is the intention of MIEMSS to start operating in the narrowband mode during the second half of calendar year 2012. Moving into narrowband operations will involve switching over EMS Operating Programs (EMSOPs) in phases over a multi-week period. It is MIEMSS' hope that all counties and/or volunteer companies will have taken advantage of the Radio Grant Program and will have replaced nonnarrowband-capable radios with narrowband-capable radio equipment by July 1, 2012.

Montana

Opportunity Title: Montana Emergency Medical Services Grant Program

Sponsor: Montana DOT.

Purpose: The Montana DOT provides competitive grants to EMS providers for acquiring or leasing ambulances or emergency-response vehicles or for purchasing equipment, other than routine supplies, for training, communication, and medical treatment.

Eligible Activities: Ambulance and emergency-response vehicle purchases, training, communications equipment, patient-care equipment.

Eligibility: A licensed EMS provider may apply if 1) it has been in operation for at least 12 months, 2) it bills for services at a level that is at least equivalent to the Medicare billing level, and 3) a majority of its active EMTs are volunteer.

Website: www.mdt.mt.gov

Contact: Montana Department of Transportation
(406) 444-7289

Notes: An eligible EMS applying for a grant under this section shall provide a 10-percent match for any grant funds.

New Mexico

Opportunity Title: New Mexico Fire Protection Grant

Sponsor: New Mexico Public Regulation Commission.

Purpose: This grant provides funding for apparatus, PPE, communications equipment, or other justified needs.

Eligible Activities: Class A or Urban-Interface pumpers, water tankers, PPE, communications and radio equipment, pump testing, wildland apparatus, and other approved expenditures.

Eligibility: Insurance Services Office (ISO) Class 7, 8, 9, or 10 fire departments.

Contact: New Mexico Fire Marshal's Office
Attention: Fire Protection Grant Application
P.O. Box 1269
Santa Fe, NM 87504
(505) 476-0174

Notes: Requires a minimum matching fund commitment of 20 percent from the department. All applicants must be in compliance with NFIRS.

North Carolina

Opportunity Title 1: Fire/Rescue Grants and Relief Fund—Volunteer Rescue and EMS Fund

Sponsor: North Carolina Office of the State Fire Marshal.

Purpose: To assist rescue organizations to purchase equipment and make capital expenditures.

Eligible Activities: Apparatus and equipment. Furniture, copiers, fax machines, manuals, land purchases, refurbish or upgrades to SCBAs, refurbish vehicles, refurbish communications equipment are typically **not** considered for matching funds.

Eligibility: Fire departments must meet the following requirements to receive the tax that is collected within their district: 1) rescue provider is all volunteer, with the exception of 10 paid positions, 2) rescue provider is recognized by the Department of Insurance as a rescue or rescue EMS provider, 3) rescue provider must meet eligibility criteria of the North Carolina Association of Rescue and EMS, 4) county recognizes organization as providing rescue or rescue/EMS or EMS, and 5) rescue provider must be certified by the Office of Emergency Medical Services to request advanced life support (ALS) equipment.

Website: www.ncdoi.com/OSFM

Contact: The Volunteer Rescue/EMS Fund
Office of State Fire Marshal
1202 Mail Service Center
Raleigh, NC 27699-1202
(919) 661-5880

Notes: This grant requires matching funds.

Opportunity Title 2: North Carolina Highway Safety Grant

Sponsor: North Carolina DOT.

Purpose: The mission of the Governor's Highway Safety Program is to plan and execute a comprehensive highway safety program for North Carolina in order to reduce the number and severity of crashes and the resulting fatalities and injuries on the State's roadways using present and potential resources available on the local, State, and national levels.

Eligible Activities: EMS—To develop and enhance methods of responding to motor-vehicle crashes and transporting and treating injured persons.

Eligibility: Fire departments and EMS agencies.

Website: www.ncdot.gov

Contact: NCDOT Governor's Highway Safety Program
(919) 733-3083

North Dakota

Opportunity Title: North Dakota Highway Safety Grant

Sponsor: North Dakota Department of Transportation (NDDOT).

Purpose: The NDDOT Traffic Safety Office receives funds through NHTSA to administer programs to reduce motor-vehicle crashes on North Dakota's roadways and the fatalities and injuries associated with these crashes.

Eligible Activities: EMS—Funds in this permissible State practice (PSP) area are passed to the North Dakota Department of Health for 1) training, testing, certification, and continuing education for EMS personnel, and 2) emergency-vehicle operations.

Eligibility: Fire departments and EMS agencies.

Website: www.dot.nd.gov

Contact: North Dakota Department of Transportation
Traffic Safety Office
608 East Boulevard Avenue
Bismarck, ND 58505-0700

Ohio

Opportunity Title 1: Fire Training Grant

Sponsor: Ohio State Fire Marshal.

Purpose: Reimbursement is available for specific fire training classes, including the cost of training manuals and student workbooks up to the maximum amount that is authorized.

Eligible Activities: Training.

Eligibility: Fire departments that provide primary fire protection to an area with a permanent population of 25,000 or less.

Website: www.com.ohio.gov/fire

Contact: (800) 515-0023

Notes: Training grant awards are subject to availability of funds.

Opportunity Title 2: Revolving Loan

Sponsor: Ohio State Fire Marshal.

Purpose: The Small Government Fire Department Services Revolving Loan Program was created by the Ohio General Assembly to assist local governments in funding certain fire department major-related expenses.

Eligible Activities: A revolving loan can be used to expedite the purchase of major firefighting, rescue, or EMS equipment. It can also be used for the construction or renovation of fire department buildings.

Eligibility: Fire departments.

Website: www.com.ohio.gov/fire

Contact: Department of Commerce, Administration–Fiscal East
P.O. Box 4009
6606 Tussing Road
Reynoldsburg, OH 43068-9009
(800) 515-0023

Opportunity Title 3: Ohio EMS Grants Program

Sponsor: Ohio EMS Board.

Purpose: The program is broken up into four different grants, identified by their priority. The first priority is given to EMS organizations for the training of personnel, for the purchase of equipment and vehicles, and to improve the availability, accessibility, and quality of EMS in Ohio. Lower priority is given to research into public education of injury prevention, effectiveness of medical procedures related to adult and pediatric trauma, and social support service mechanisms for trauma victims and their families.

Eligible Activities: Training, purchase of equipment and vehicles, trauma-care research.

Eligibility: Public, private, and nonprofit EMS agencies and EMS agencies established by a joint fire district.

Website: www.ems.ohio.gov

Contact: Grants Administration
Ohio Division of EMS
1970 West Broad Street
Columbus, OH 43223
(800) 233-0785

Opportunity Title 4: Rural Community Financial Assistance (RCFA)

Sponsor: Ohio State Fire Marshal.

Purpose: The RCFA is a matching program to cover the cost of tuition and lodging at the Division of State Fire Marshal's Ohio Fire Academy. Recipients are responsible for travel, meals, and other expenses. Only firefighters from communities serving a population of less than 10,000 qualified. However, for multicommunity projects, the 10,000 may be exceeded, provided none of the communities in the project have more than 10,000 people.

Eligible Activities: Training at Ohio Fire Academy.

Eligibility: Fire departments serving a population under 10,000.

Website: www.com.ohio.gov/fire

Contact: (888) 726-7731

Notes: This fund will not cover cost incurred for State certification (fire or EMS) training.

Oklahoma

Opportunity Title 1: Rural Fire 80/20 Reimbursement Grant

Sponsor: Oklahoma Division of Forestry.

Purpose: The objective of this program is to provide financial assistance to rural fire departments serving communities under 10,000 in population. With this grant, communities can purchase equipment for the fire department or build, remodel, or expand a fire station.

Eligible Activities: Fire-station construction and purchase of equipment.

Eligibility: Fire departments or fire districts serving a population under 10,000.

Website: www.forestry.ok.gov

Contact: Oklahoma Department of Agriculture, Food, and forestry
2800 N. Lincoln Boulevard
P.O. Box 528804
Oklahoma City, OK 73152
(405) 288-2385

Notes: No advanced payments will be made. Grant amounts may be claimed only on a reimbursement basis.

Opportunity Title 2: Community Wildfire Protection Plan (CWPP) Reimbursement Grant

Sponsor: Oklahoma Division of Forestry.

Purpose: The grant reimburses costs incurred for the development of a CWPP in accordance with minimum requirements for such plans.

Eligible Activities: Wildfire planning.

Eligibility: Eligible applicants are communities (cities, towns) and fire departments. County governments are not eligible.

Website: www.forestry.ok.gov

Contact: Oklahoma Department of Agriculture, Food, and Forestry
2800 North Lincoln Boulevard
P.O. Box 528804
Oklahoma City, OK 73152
(405) 288-2385

Notes: This is a 100-percent Federal passthrough reimbursement grant. Recipients can only make purchases after their community has been officially awarded a grant and received a copy of the State purchase order. Expenses prior to the State purchase order date will not be eligible for this grant.

Oregon

Opportunity Title: Oregon Highway Safety Program

Sponsor: Oregon DOT.

Purpose: To provide funding to community hospitals and/or EMS providers for community hospital projects that affect the treatment and outcome of traffic-related injuries.

Eligible Activities: EMS—for training, equipment, and outreach.

Eligibility: Applicants must be community hospitals or their EMS providers.

Website: www.oregon.gov/ODOT/TS/safetyprograms.shtml

Contact: Oregon Department of Transportation
Transportation Safety Division
EMS Program Manager
235 Union Street Northeast
Salem, OR 97301-1054
(503) 986-3293

Pennsylvania

Opportunity Title 1: Volunteer Fire Company and Volunteer Ambulance Service Grant Program (VFCVASGP)

Sponsor: Pennsylvania Office of the State Fire Commissioner.

Purpose: The VFCVASGP provides grants for volunteer fire, ambulance, and rescue companies.

Eligible Activities: Construction and/or renovation of facilities; purchase or repair of equipment; debt reduction associated with facilities or equipment purchases; and training and certification of members.

Eligibility: Volunteer fire companies, volunteer ambulance services, and volunteer rescue squads.

Website: www.portal.state.pa.us/portal/server.pt/community/state_fire_commissioner_home/4462

Contact: (717) 651-2207

Opportunity Title 2: Emergency Medical Services Operating Fund (EMSOF)

Sponsor: Pennsylvania Department of Health.

Purpose: To provide funding to maintain, improve, and develop the quality of the EMS system within this Commonwealth.

Eligibility: Any provider of EMS applying for EMSOF funding must be in full compliance with all regulations, policies, and priorities of the State and regional EMS systems.

Website: www.dsf.health.state.pa.us/health/lib/health/ems/emsof.pdf

Opportunity Title 3: Volunteer Loan Assistance Program (VLAP)

Sponsor: Pennsylvania Office of the State Fire Commissioner (OSFC).

Purpose: The VLAP provides **low-interest loans** to volunteer fire companies, ambulance services, and rescue squads for the purpose of purchasing and modernizing apparatus, facilities, and equipment. Forms and informational materials are available to interested organizations. OSFC staff provides technical assistance and regulatory guidance to applicants; receives, analyzes, and approves loan applications; supplies loan-closing services; maintains financial and historical files and services all accounts until they are closed.

Eligibility: All Pennsylvania volunteer fire companies, ambulance services, and rescue squads.

Website: www.portal.state.pa.us/portal/server.pt/community/state_fire_commissioner_home/4462/loans_and_grants/567304

Tennessee

Opportunity Title: Tennessee Governor's Highway Safety Office Grants

Sponsor: Tennessee DOT.

Purpose: The Governor's Highway Safety Office provides grants to programs which are designed to reduce the number of fatalities, injuries, and related economic losses resulting from traffic crashes on Tennessee's roadways. Local governments, law enforcement agencies, academic institutions, and private nonprofits can apply for NHSTA passthrough funding for projects related to various areas of highway safety.

Eligible Activities: Eligible project areas are included in the State of Tennessee's Highway Safety Performance Plan. The plan includes EMS.

Eligibility: Fire departments and EMS agencies.

Website: www.tdot.state.tn.us/ghso/grants.htm

Contact: Governor's Highway Safety Office
(615) 687-2405

Texas

Opportunity Title 1: EMS Local Projects Grant (LPG)

Sponsor: Texas Department of State Health Services.

Purpose: The purpose of the LPG program is to support and improve the development of the Texas Emergency Health Care System. Funds shall be used to support and improve the development of the Texas Emergency Health Care System and increase the availability and quality of emergency prehospital health care.

Eligible Activities: EMS personnel certification training, specialty training related to prehospital healthcare management, communication equipment, patient-care equipment and nondisposable supplies, injury-prevention projects, continuing education programs, and the purchase of an ambulance.

Eligibility: Licensed EMS providers, registered first-responder organizations, Regional EMS/Trauma Advisory Councils (RACs), EMS education organizations, and prehospital injury-prevention organizations.

Website: www.dshs.state.tx.us/default.shtm

Contact: Texas Department of State Health Services
Office of Emergency Medical Services/Trauma Systems Coordination
(512) 458-7470

Notes: Matching funds requirement.

Opportunity Title 2: Governor's Extraordinary Emergency Fund

Sponsor: Texas Department of State Health Services.

Purpose: Funding is set aside each fiscal year from the EMS and Trauma Care System Account (911 Funds), the Trauma Facilities and Trauma Care System Fund (enacted by Senate Bill 1131, 78th Legislature), and the Designated Trauma Facility and Emergency Medical Services Account (enacted by House Bill 3588, 78th Legislature) to support the emergent, unexpected needs of EMS providers or Department of Social and Health Services (DSHS)-approved organizations. Proposals are evaluated based on impact to the regional or Statewide EMS/trauma system.

Eligible Activities: Funding for emergent, unanticipated needs of EMS providers. Emergency funds are not intended to expand service.

Eligibility: Licensed EMS providers, registered first-responder organizations, and licensed hospitals.

Website: www.dshs.state.tx.us/default.shtm

Contact: Texas Department of State Health Services
Office of EMS/Trauma Services
MC 1876, P.O. Box 149347
Austin, TX 78714-9347
(512) 834-6700, Extension 2377

Utah

Opportunity Title: Utah Fire Department Assistance Grant

Sponsor: Utah Division of Forestry, Fire and State Lands.

Purpose: The fire department assistance grant is an interagency, cooperative program, administered by the Division of Forestry, Fire & State Land (FFSL), to provide financial assistance to Utah's fire departments. Funding can be used for wildland or structural fire training, prevention, PPE, and basic suppression equipment. This grant is made up of three different sources of funding; two Federal and one State; RFA from the DOI, VFA from the U.S. Forest Service, and funds through the Utah Fire and Rescue Academy (UFRA) made available by the Utah Fire Marshal Office through the Fire Prevention Board. These funding sources have been blended into a single program in an attempt to simplify the process for Utah's fire departments.

Eligible Activities: Initial organizations (start-up costs), training, fire shelters, fire prevention equipment and materials, PPE, basic suppression tools and equipment, and other special needs.

Eligibility: Fire departments.

Website: www.ffsl.utah.gov

Contact: Utah Division of Forestry, Fire and State Lands
1594 West North Temple, Suite 3520
Salt Lake City, UT 84114-5703
(801) 538-5427

Notes: All grant awards require a match from the recipient. VFA money is a 50/50 match. RFA and UFRA monies are a 90/10 match.

Virginia

Opportunity Title 1: Virginia Fire Incident Reporting System (VFIRS) Grant

Sponsor: Virginia Department of Fire Programs (VDFP).

Purpose: The purpose of this grant program is to provide an incentive to those jurisdictions not currently participating in VFIRS by lessening the economic burden.

Eligible Activities: VFIRS 'hardware'—inclusive of a PC, printer, and connectivity equipment.

Eligibility: Any independent city, county, or incorporated town is eligible to apply.

Website: www.vafire.com

Contact: Information Statistics Manager
Virginia Department of Fire Programs
1005 Technology Park Drive
Glen Allen, VA 23059-4500
(804) 249-1990

Opportunity Title 2: Fire Service Board Mini Grant

Sponsor: VDFP.

Purpose: The objective of the Virginia Fire Services Board (VFSB) Training Mini-Grant program shall be 1) to enhance fire training facilities and capabilities, 2) develop application and program processes, 3) offer equal opportunity for localities to apply and be considered, 4) assist VFSB in making fair awards, and 5) permit VDFP to efficiently administer the program.

Eligible Activities: Training projects.

Eligibility: Independent cities, counties, or towns incorporated within counties. Application must be made through a locality's government.

Website: www.vafire.com

Contact: Grants Manager
1005 Technology Park Drive
Glen Allen, VA 23059-4500
(804) 371-0220

Notes: No single locality (city, town, or county) shall make application for nor be eligible for more than one award per Mini-Grant cycle.

Opportunity Title 3: Rescue Squad Assistance Fund

Sponsor: Virginia Department of Health.

Purpose: The Financial Assistance for Emergency Medical Services Grants Program, known as the Rescue Squad Assistance Fund (RSAF) Grant Program, is a multimillion dollar grant program for Virginia nonprofit EMS agencies and organizations.

Eligible Activities: Items eligible for funding include EMS equipment and vehicles, computers, EMS management programs, courses/classes and projects benefiting the recruitment, and retention of EMS members.

Eligibility: Virginia nonprofit agency/organization involved in EMS.

Website: www.vdh.virginia.gov/OEMS/Grants/index.htm

Contact: Grants Manager
Office of Emergency Medical Services
1041 Technology Park Drive
Glen Allen, VA 23059-4500
(804) 964-7600

Opportunity Title 4: EMS Training Funds Program

Sponsor: Virginia Office of EMS.

Purpose: The EMS Training Fund program is designed to provide financial assistance for Virginia-certified EMS providers and Virginia Office of EMS-approved EMS courses. These funds shall supplement local support for EMS courses.

Eligible Activities: Virginia Office of EMS-approved EMS courses.

Eligibility: Nonprofit entities and Virginia-certified EMS providers.

Website: www.vdh.state.va.us/OEMS/Training/EMSTF.htm

Contact: Virginia Office of EMS
1041 Technology Park Drive
Glen Allen, VA 23059

Opportunity Title 5: Conference and Education Assistance Program

Sponsor: VDFP.

Purpose: The purpose of the Conference and Education Assistance Program is to provide needed financial support for conferences and seminars sponsored by the Virginia-based non-profit organizations that further the education of fire and emergency services personnel throughout the Commonwealth.

Eligible Activities: Conferences and seminars, fire prevention and public safety education courses.

Eligibility: Virginia-based, legally-recognized nonprofit organizations representing the local, regional, or Statewide interests of the Commonwealth's fire and emergency services community.

Website: www.vafire.com

Contact: Grants and Local Aid Manager
Virginia Department of Fire Programs
1005 Technology Park Drive
Glen Allen, VA 23059-4500
(804) 371-0220

Opportunity Title 6: Virginia Fire Programs Fund—Aid to Localities

Sponsor: VDFP.

Purpose: The Virginia Fire Programs Fund is derived from 1 percent of fire-related insurance coverage. Approximately 75 percent of the total fund goes directly to counties, cities, and incorporated towns within the Commonwealth as "Aid to Localities."

Eligible Activities: Training, construction of training centers, firefighting equipment, and PPE.

Eligibility: Cities, counties, and incorporated towns.

Website: www.vafire.com

Contact: Grants and Local Aid Manager
Virginia Department of Fire Programs
1005 Technology Park Drive
Glen Allen, VA 23059-4500
(804) 249-1974

Notes: Allocations are population-based.

Opportunity Title 7: Burn Building Grants

Sponsor: VDFP.

Purpose: Burn Building grants are provided for the construction, renovation, or repair of burn buildings in the Commonwealth of Virginia.

Eligible Activities: Construction, renovation, or repair of burn buildings.

Eligibility: Localities within the Commonwealth of Virginia otherwise eligible to receive "Aid to Localities" from the "Fire Programs Fund."

Website: www.vafire.com

Contact: Grants and Local Aid Manager
Virginia Department of Fire Programs
1005 Technology Park Drive
Glen Allen, VA 23059-4500
(804) 371-0220

Notes: The Burn Building Grant Program for permanent burn buildings is a matching/cost-sharing program.

Washington

Opportunity Title 1: Wildland Fire Training

Sponsor: Washington State Department of Natural Resources (DNR).

Purpose: DNR sponsors wildland fire training in partnership with Wenatchee Valley College and with support from the U.S. DOI.

Eligible Activities: Wildland firefighting courses.

Eligibility: Rural fire departments and fire districts.

Website: www.dnr.wa.gov/RecreationEducation/Topics/FireInformation/Pages/rp_fire_dnrfdassistprog.aspx

Contact: DNR Fire District Assistance Coordinator
(360) 902-1306

Opportunity Title 2: Wildland Fire Assistance Grants

Sponsor: Washington State DNR.

Purpose: Administered by DNR through funding from the U.S. Department of Agriculture, this grant program provides 50-percent match for purchases of PPE and general equipment.

Eligible Activities: PPE and general equipment.

Eligibility: Opportunities for these grants are available to fire protection districts and departments in Washington State that 1) respond to wildland fires on private, State, or Federal lands, 2) serve communities with a population of 10,000 or fewer residents, and 3) serve a community of more than 10,000 residents **and** a service area that includes a rural community of fewer than 10,000 residents.

Website: www.dnr.wa.gov/RecreationEducation/Topics/FireInformation/Pages/rp_fire_dnrfdassistprog.aspx

Contact: DNR Fire District Assistance Coordinator
(360) 902-1306

Notes: A 50-percent match is required.

Opportunity Title 3: Firefighter Property Program

Sponsor: Washington State DNR.

Purpose: This program helps fire protection districts and fire departments get fire engines and fire tenders (vehicles that bring water, foam, or dry chemicals to fire trucks in the field) suitable for low-cost conversion to wildland use. Fire districts receive the title to the property. Districts may have to pay the expense of transporting the vehicle from an out-of-State location.

Eligible Activities: Wildland fire apparatus.

Eligibility: Opportunities to obtain equipment through this program may be available to fire protection districts and fire departments in Washington that 1) agree to put the vehicle in service within a year of taking possession, and 2) are willing to assist DNR in protecting Washington wildlands from wildfire.

Website: www.dnr.wa.gov/RecreationEducation/Topics/FireInformation/Pages/rp_fire_dnrfdassistprog.aspx

Contact: DNR Fire District Assistance Coordinator
(360) 902-1306

Notes: This program replaces the Federal Excess Property Program (FEPP) in Washington State.

Wisconsin

Opportunity Title 1: Preparedness and Volunteer Training

Sponsor: Wisconsin Office of Justice Assistance.

Purpose: This grant program is designed to enhance community emergency preparedness and participation capabilities by engaging citizens to become fully aware, trained, and practiced on how to prevent, protect/mitigate, prepare for, and respond to all threats and hazards. Grant funds will enable training for personal preparedness, instructors, exercises, ongoing volunteer programs, and surge capacity response.

Eligible Activities: Funding may be used for personnel, employee benefits, travel/training, supplies, and other operating expenses, consultants/contractual and other.

Eligibility: Open to qualified units of local government on a competitive basis.

Website: www.ojp.usdoj.gov/saa/wi.htm

Contact: Wisconsin Office of Justice Assistance
1 South Pinckney Street, Suite 615
Madison, WI 53703-3220
(608) 261-7538

Notes: All expenses must be new and cannot replace existing State or local government funding.

Opportunity Title 2: National Incident Management System (NIMS) and Incident Command System (ICS) Training Program

Sponsor: Wisconsin Office of Justice Programs.

Purpose: Funding under this grant program will help local and State government agencies become compliant with NIMS and ICS requirements. Compliance with the Federal requirements will improve the State's ability to respond to all-hazards emergencies whether it is local, regional, or Statewide.

Eligible Activities: Funding may be used to reimburse costs associated with the hosting of eligible NIMS and ICS training classes. Eligible costs include instructor fees/expenses, student supplies, facilities, and training props.

Eligibility: Open to qualified units of local government on a competitive basis. Must comply with Homeland Security Presidential Directive 5 (HSPD-5).

Website: www.ojp.usdoj.gov/saa/wi.htm

Contact: Wisconsin Office of Justice Assistance
1 South Pinckney Street, Suite 615
Madison, WI 53703-3220
(608) 261-7538

Notes: All expenses must be new and cannot replace existing State or local government funding.

Opportunity Title 3: Homeland Security Exercise Grant

Sponsor: Wisconsin Office of Justice Assistance.

Purpose: This grant program provides funds to design, develop, conduct, and evaluate exercises to test the plans and capabilities of Wisconsin's emergency-response community.

Eligible Activities: Allowable exercise costs include expenses related to event supplies, rental of equipment and/or facilities, travel, backfill, and overtime for participants and event planners.

Eligibility: Open to qualified units of local government on a competitive basis.

Website: www.ojp.usdoj.gov/saa/wi.htm

Contact: Wisconsin Office of Justice Assistance
1 South Pinckney Street, Suite 615
Madison, WI 53703-3220
(608) 261-7538

Notes: All expenses must be new and cannot replace existing State or local government funding.

Opportunity Title 4: Forest Fire Protection (FFP) Grant

Sponsor: Wisconsin DNR.

Purpose: FFP grants are available to Wisconsin fire departments and county/area fire associations. Grant funding is intended to expand the use of local fire departments to augment and strengthen the DNR overall initial-attack fire suppression capabilities on forest fires.

Eligible Activities: PPE, forest-fire training; forest-fire prevention and WUI; forest-fire suppression tools, equipment, supplies and materials; dry hydrants; communication equipment for forest-fire suppression or protection; mapping and GIS; and off-road, all-wheel drive vehicles, all-terrain vehicles of 1/2-ton to 5-ton capacity that are used primarily for forest-fire suppression.

Eligibility: Wisconsin fire departments that have an executed forest-fire suppression agreement with DNR. Wisconsin county/area fire associations if a majority of the fire department members serve organized forest-fire control areas designated by the DNR.

Website: www.dnr.wi.gov/org/caer/cfa/LR/ffp/grants.html

Contact: Grant Manager
Wisconsin Department of Natural Resources
P.O. Box 7921
Madison, WI 53707-7921
(608) 267-0848

Notes: The FFP program is a 50/50 cost-share reimbursement grant program.

Wyoming

Opportunity Title 1: Mineral Royalty Grants

Sponsor: Wyoming Office of State Lands and Investments.

Purpose: Pursuant to W.S. 9-4-604(a), the Mineral Royalty Grants are awarded to 1) alleviate an emergency situation which poses a direct and immediate threat to health, safety, or welfare, or 2) to comply with a Federal or State mandate, or 3) to provide an essential public service.

Eligible Activities: Varied, includes water projects, acquisition of emergency vehicles, public administration buildings, and health-care facilities, among others.

Eligibility: Counties, municipalities, Joint Powers Boards, and certain special districts.

Website: http://slf-web.state.wy.us/grants/mrgsheet.aspx

Contact: Office of State Lands and Investments
122 West 25th Street
Cheyenne, WY 82002

Opportunity Title 2: Wyoming Traffic Safety Grant

Sponsor: Wyoming DOT.

Purpose: The Wyoming Department of Transportation's Highway Safety Grant Program's goal is to reduce traffic crashes, deaths, injuries, and the associated economic losses in Wyoming through the coordinated efforts of State, local agencies, and other safety partners. The Highway Safety Program provides project funding to help achieve stated programs' goals and objectives which address identified highway safety problems. Priorities of safety effort and location are based on projects with the greatest likelihood to impact crash and injury reduction.

Eligible Activities:	EMS. Under the statutory provisions administered by the Federal Highway Administration (FHWA), the safety areas.
Eligibility:	Fire departments and EMS agencies.
Website:	www.dot.state.wy.us
Contact:	Highway Safety Program Office Wyoming Department of Transportation 5300 Bishop Boulevard Cheyenne, WY 83009-3340 (307) 777-4257

Chapter 6: Federal Funding for Emergency Medical Services and Fire Agencies

This chapter lists major grants available from key Federal agencies that can be used to secure alternative funding for fire protection, emergency medical services (EMS), and disaster preparedness and recovery.

The Federal government is the largest source of grants for the EMS and fire services. There are more than 1,000 grant programs offered by 26 Federal grant-making agencies. Since 9/11, billions of Federal dollars have been made available to States, local fire departments, and EMS organizations in the form of grants and funding programs for homeland security and related programs, including EMS.

Much of the Federal grant budget is passed to the States through formula or block grants. Examples include the Emergency Management Performance Grants (EMPG) and the Homeland Security Grant Program (HSGP). From there, it is up to the States to decide how to use the money. There are, however, direct Federal grant programs to fire and EMS agencies such as the Assistance to Firefighters Grant (AFG).

The State governor's office is usually an excellent resource for information on Federal funds administered by the State. Local members of the U.S. House of Representatives or the U.S. Senate also may be helpful in bridging the gap between Federal programs and local governments or nonprofit EMS agencies.

Federal-Funding Basics

Nearly every Federal government agency has its own grant or loan program. Each is intended to serve a particular purpose and comes with its own set of rules and program guidance.

Understanding the various types of Federal grants is important because the funding mechanism selected will influence the strategy employed to access the funds and how funds can be used. For example, many of the grants identified here are competitive grants. A competitive grant, or project grant, is one where applicants vie for limited funds. Peer-reviewers score applications and money is awarded to those applicants with the highest scores. Federal grants may be direct or passthrough. Direct grants are given directly to the agency applying for it; whereas passthrough grants require the State to apply to the Federal government, then the State hands out grant money to agencies that request it. Grants may be unrestrictive in terms of use, but usually come with specific guidelines stated in the program guidance.

This guide focuses on grants, but Federal loan programs are also available as an alternative-funding source. Loan funds go directly to the applicant, which is responsible for repayment. The main advantage of a grant is that it does not have to be paid back as long as the conditions of the grant program are met.

Types of Federal Grants

Federal grants can be grouped into the following major categories.

Block Grants

A block grant does not involve competition. The Federal government simply distributes funds to the States based on an established formula. Formula grants flow directly to State agencies that subgrant the funds through a proposal process or otherwise turn the funds over to local governments or nonprofit agencies.

Project Grants

Project grants are the most common form of Federal grant. Depending on the program requirements, EMS organizations gain access to the funds through a competitive-bidding process. Application to a project grant does not guarantee an award and the amount received by grantees is not predetermined by a formula.

Demonstration Grants

Demonstration grants are pilot projects generally involving a small number of sites in an effort to learn more about the effectiveness of a new program. An effective demonstration grant program may lead to further funding in the form of discretionary or project grants. Demonstration grants are awarded competitively and can go to State or local governments or community-based organizations depending on the eligibility requirements.

Congressional Earmarks

Earmarks are explicitly specified in appropriations by the U.S. Congress. They are not competitively awarded and have become highly controversial because of the role of paid political lobbyists in securing them.

Data Universal Numbering System Number

All organizations applying for a Federal grant or cooperative agreement must have a Data Universal Numbering System (DUNS) number. Individuals who would personally receive a grant or cooperative agreement award from the Federal government apart from any business or nonprofit organization they may operate, and foreign entities, are exempt from this requirement.

The DUNS number is a unique nine-character identification number provided by the commercial company Dun & Bradstreet (D&B). The DUNS number is D&B's copyrighted, proprietary means of identifying business entities on a location-specific basis worldwide.

A DUNS number remains with the company location to which it has been assigned even if it closes or goes out-of-business. The DUNS number is widely used by both commercial and Federal entities and was adopted as the standard business identifier for Federal electronic commerce in October 1994. The DUNS was also incorporated into the Federal Acquisition Regulation (FAR) in April 1998 as the Federal government's contractor identification code for all procurement-related activities.

DUNS Q&A: www.usda.gov/rus/telecom/dlt/pdf_files/duns_qa.pdf

DUNS Number Guide: www.ccr.gov/pdfs/DUNSGuideGovVendors.pdf

Request a DUNS number by web at: http://fedgov.dnb.com/webform or toll free at (866) 705-5711

Federal Grants

U.S. Department of Health and Human Services (HHS)

Opportunity Title 1: Health Resources and Services Administration (HRSA) Rural Health Care Service Outreach Grant

Purpose:	The Office of Rural Health Policy's Rural Health Care Services Outreach Grant Program encourages the development of new and innovative health-care delivery systems in rural communities that lack essential health-care services. The emphasis of this grant program is on service delivery through collaboration, requiring the lead applicant organization to form a consortium with at least two additional partners. The community being served must be involved in the development and ongoing operations of the program to appropriately address the needs of the population.
Eligible Activities:	Rural medical services.

Eligibility: Eligible lead applicants are rural, nonprofit private or public entities that represent a consortium of three or more organizations that deliver health-care services in rural areas. Only the applicant organization is required to be a rural private or public not-for-profit entity. Faith-based and community-based organizations are eligible to apply.

Website: www.ruralhealth.hrsa.gov/funding/outreach.htm

Contact: Outreach Grant Program Coordinator
HRSA, Office of Rural Health Policy
(301) 443-6894
Fax: (301) 443-2803

Opportunity Title 2: Delta Health Initiative Cooperative Agreement Program

Purpose: The Delta Health Initiative Cooperative provides funding to an alliance to address long-standing, unmet rural health needs of the Mississippi Delta Region.

Eligible Activities: Projects funded under this cooperative agreement may include, but are not limited to, improving communication and coordination among key health-care providers; implementing health education, intervention, and wellness promotion; improving access to health-care services (e.g., nonemergency medical transport); improving quality health assessment data to promote information-based health services; education and health literacy; workforce training and entrepreneurial opportunities; telehealth and other health information technologies; and new construction of public health-related facilities.

Eligibility: Organizations must be a non-Federal, not-for-profit alliance. The alliance must consist of no fewer than four academic institutions and must include the State Medical Association and State Hospital Association.

Website: www.hhs.gov

Contact: Delta Health Initiative Program Coordinator
DHHS/HRSA, Office of Rural Health Policy
(301) 443-7321
Fax: (301) 443-2803

Notes: HRSA is requiring applicants for this funding opportunity to apply electronically through Grants.gov All applicants must submit in this manner unless the applicant is granted a written exemption from this requirement in advance by the Director of HRSA's Division of Grants Policy.

Opportunity Title 3: Rural Access to Emergency Devices Grant Program

Purpose: The purpose of the Rural Access to Emergency Devices Grant Program is to purchase automated external defibrillators (AEDs) that have been approved or cleared for marketing by the Food and Drug Administration (FDA) and to provide defibrillator and basic life support (BLS) training in AED usage through the American Heart Association, the Red Cross, or other nationally-recognized training courses.

Eligible Activities: Purchase of AED devices.

Eligibility: Awards will be made to community partnerships. These partnerships are defined as a consortium of first responders (e.g., EMS, law enforcement, and fire departments) and local for-profit and nonprofit entities that may include, but are not limited to, long-term care facilities, rural-health clinics, community-health centers, post offices, libraries, and other civic centers, athletic facilities, and senior organizations applying as a community partnership. All applicant organizations have to be located in an eligible rural county or eligible rural census tract of urban counties.

Website: http://ruralhealth.hrsa.gov/funding/aed.htm Eligible rural counties can be found at: http://ruralhealth.hrsa.gov/funding/eligibility The eligible census tracts of urban counties are included in the document identified above. To identify the Census Tract where your organization is located, visit the webpage at: http://app.ffiec.gov/geocode/default.htm

Contact: RAED Program Coordinator
Office of Rural Health Policy
(301) 443-7529
Fax: (301) 443-2803

Opportunity Title 4: Public Access to Defibrillation Demonstration Projects (PADDP)

Purpose: The purpose of the PADDP grant program is to develop and implement innovative, comprehensive, community-based public-access defibrillation demonstration projects that: 1) provide cardiopulmonary resuscitation (CPR) and automated external defibrillation to cardiac arrest victims in unique settings, 2) provide training to community members in CPR and automated external defibrillation, and 3) maximize community access to AED.

Eligible Activities: Purchase of AED devices and associated Public Access Defibrillation projects.

Eligibility: To be eligible, the applicant must be a political subdivision of a State, a Federally-recognized Native American community, or a tribal organization. Partnerships may be composed of emergency-response entities such as training facilities, local emergency responders, fire and rescue departments, police, community hospitals, and nonprofit entities and for-profit entities concerned about cardiac-arrest survival rates.

Website: www.hrsa.gov/ruralhealth/about/community/defilibration.html

Contact: PADDP Program Coordinator
Office of Rural Health Policy-Room 9A-55
5600 Fishers Lane
Rockville, MD 20857
(301) 443-7320
Fax: (301) 443-2803

Opportunity Title 5: Small Rural Hospital Improvement Program (SHIP)

Purpose: The purpose of this grant is to help small, rural hospitals to 1) pay the costs related to implementation of prospective payment systems (PPS), 2) comply with provisions of the Health Insurance Portability and Accounting Act (HIPPA) of 1996, and 3) reduce medical errors and support quality improvement.

Eligible Activities: Grants may be used to purchase technical assistance, services, training, and information technology. Proposed initiatives should include efforts to support quality improvement and adopting of health-information technology.

Eligibility: The SHIP grant program funds are geared towards assisting small, rural hospitals that are essential access points for Medicare and Medicaid beneficiaries. Eligible small, rural hospitals are non-Federal, short-term general acute-care facilities that are located in a rural area of the United States and the territories, including faith-based hospitals. For the purpose of this program, 1) small is defined as 49-staffed beds or less and 2) rural is defined as either located outside of a Metropolitan Statistical Area (MSA) or located within a rural census tract of a MSA, as determined under the Goldsmith Modification or the Rural Urban Commuting Areas (RUCAs). Hospitals may be for-profit or not-for-profit. Tribally-operated hospitals under Titles I and V of P.L. 93-638 are eligible to the extent that such hospitals meet the above criteria. Regardless of geographic location, all designated Critical Access Hospitals (CAHs) are eligible.

Website: A link to the application is available through Grants.gov at: https://apply07.grants.gov/apply/UpdateOffer?id=17401

Contact: SHIP Program Coordinator
Office of Rural Health Policy
(301) 443-0835
Fax: (301) 443-2803

Notes: Eligible hospitals should contact their State Office of Rural Health.

Opportunity Title 6: Small Healthcare Provider Quality Improvement (SHCPQI) Program

Purpose: The purpose of the SHCPQI grant is to assist rural providers with the implementation of quality improvement strategies, while improving patient care and chronic disease outcomes. The focus of the SHCPQI grant is on quality improvement for the following chronic diseases: diabetes mellitus (DM) and cardiovascular disease (CVD).

Eligible Activities: Primary care quality improvement programs.

Eligibility: To be eligible for SHCPQI, applicants must meet one of the following criteria: 1) be located in rural areas as determined by eligible rural county census tracts, 2) the applicant exists exclusively to provide services to migrant and seasonal farm workers in rural areas, or 3) be a Tribal government whose grant-funded activities will be conducted within their Federally-recognized Tribal area.

Contact: SHCPQI Program Coordinator
(301) 443-4107
Fax: (301) 443-2803

Opportunity Title 7: Delta Rural Hospital Performance Improvement (RHPI) Project

Purpose: The RHPI project focuses on strengthening rural hospitals as the cornerstone to preserving health-care access.

Eligible Activities: Performance-improvement activities.

Eligibility: Eligible hospitals are those with fewer than 200 beds and are located in the Delta Regions of Alabama, Arkansas, Illinois, Kentucky, Louisiana, Mississippi, Missouri, and Tennessee. Mountain States Group (MSG) works in coordination with the State Offices of Rural Health, the State Hospital Associations, and quality improvement organizations to coordinate activities. In addition to size and location, there are other selection criteria for individual telehealth advancement (TA).

Website: www.hrsa.gov/ruralhealth/about/delta/

Contact: Delta RHPI Program Manager
Mountain States Group
1607 W. Jefferson
Boise, ID 83702-5111
(904) 553-0081

Opportunity Title 8: School-Based Health Center Capital (SBHCC) Program

Purpose: SBHCC grants will address significant and pressing capital needs to improve delivery and support expansion of services at school-based health centers.

Eligible Activities: Telehealth equipment; equipment to support an electronic health record (EHR) system; medical equipment; office equipment; school health-center construction.

Eligibility: An eligible applicant must be a school-based health center or a sponsoring facility of a school-based health center as defined in Section 4101(a)(6) of the Affordable Care Act, as set forth in Section 2110(c)(9) of the Social Security Act (42 USC 1397jj(c)(9)).

Website: www.hss.gov

Contact: Office of Policy and Program Development
(301) 594-4300

Opportunity Title 9: Rural Health Network Development (RHND) Grant Program

Purpose: The purpose of the RHND grant program is to expand access to, coordinate, and improve the quality of essential health-care services, and enhance the delivery of health care in rural areas. These grants support rural providers who work in formal networks, alliances, coalitions, or partnerships to integrate administrative, clinical, technological, and financial functions. Funds provided through this program are not used for direct delivery of services. The ultimate goal is to strengthen the rural health-care delivery system by 1) improving the viability of the individual providers in the network, and/or 2) improving the delivery of care to people served by the network.

Eligible Activities: Rural health-care services.

Eligibility: The applicant organization must be a public or nonprofit entity that is a network or is a member of a network that includes at least three separately owned health-care providers or other entities that provide or support the delivery of health-care services.

Contact: Program Coordinator
Office of Rural Health Policy
(301) 443-8335
Fax: (301) 443-2803

Notes: A funding preference will be given to qualified applicants that can demonstrate any of the following three criteria: 1) Those applicants where the service area is located in an officially-designated health professional shortage area (HPSA), or 2) is a medically-underserved community (MUC) or serves medically-underserved populations (MUPs), or 3) applicants whose projects focus on primary care and/or wellness and prevention strategies.

U.S. Department of Agriculture (USDA)

Opportunity Title 1: Rural Development Community Facilities Program

Purpose: Community programs provide loans and grants and loan guarantees for water and environmental projects, as well as community-facilities projects. Water and environmental projects include water systems, waste systems, solid waste, and storm-drainage facilities. Community facilities projects develop essential community facilities for public use in rural areas and may include hospitals, fire protection, safety, as well as many other community-based initiatives.

Eligible Activities: Hospitals, fire protection, safety, EMS, ambulances.

Eligibility: Rural communities.

Website: www.rurdev.usda.gov/HCF_CF.html

Contact: For more information about this program, or to file an application, contact the local USDA service center in your area. The website to find your local office is: http://offices.sc.egov.usda.gov/locator/app

Opportunity Title 2: Rural Emergency Response Initiative

Purpose: To develop the capacity and ability of private, nonprofit, community-based housing and community development organizations, and low-income rural communities to improve housing, community facilities, and community and economic development projects in rural areas.

Eligible Activities: Rural Community Development Initiative grants may be used for, but are not limited to, 1) training subgrantees to conduct a program on home-ownership education; 2) training subgrantees to conduct a program for minority business entrepreneurs; 3) providing technical assistance to subgrantees on how to effectively prepare a strategic plan; 4) provide technical assistance to subgrantees on how to access alternative-funding sources; 5) building organizational capacity through board training; 6) developing training tools, such as videos, workbooks, and reference guides to be used by the subgrantee; 7) providing technical assistance and training on how to develop successful childcare facilities; and 8) providing training on effective fundraising techniques.

Eligibility: Purchase of construction of facilities including, but not limited to, fire apparatus, fire department buildings, multiservice buildings, rescue and ambulance-service buildings, rescue and ambulance and equipment, architectural and engineering feeds, and right-of-way assessments.

Website: www.rurdev.usda.gov/HAD-RCDI_Grants.html

Contact: For more information about this program, or to file an application, contact the local rural development office in your area.

Opportunity Title 3: Distance Learning and Telemedicine (DLT) Program

Purpose: The DLT program is specifically designed to meet the educational and health-care needs of rural America through the use of advanced telecommunications technologies. The DLT program provides grants, loans, and loan-grant combinations as a source of financing and technical assistance for rural telecommunications systems.

Eligible Activities: Telemedicine and telecommunications.

Eligibility: 1) Currently deliver or propose to deliver DLT services for the term of the grant. To receive a grant, the purposes must meet the grant definition of distance learning and/or telemedicine. 2) Be legally organized as an incorporated organization or partnership; an Indian tribe or tribal organization; a State or local unit of government; a consortium; or other legal entity, including a private corporation organized on a for-profit or not-for-profit basis with the legal capacity to contract with the U.S. government. 3) Operate a rural-community facility or deliver DLT services to entities that operate a rural-community facility or the residents of rural areas at rates calculated to ensure that the benefit of the financial assistance passes through such entities or to residents of rural areas.

Website: www.usda.gov/wps/portal/usda/usdahome

Contact: Director, ASD–Telecommunications Programs
STOP 1550, Room 2844
1400 Independence Avenue, Southwest
Washington, DC 20250
(202) 720-0413
Fax: (202) 720-1051

U.S. Department of Health Resources and Services Administration (HRSA)

Opportunity Title: Medicare Rural Hospital Flexibility Grant (Flex)

Purpose: The Flex grant provides funding to State governments to spur quality and performance-improvement activities; stabilize rural hospital finance; and integrate EMS into their health-care systems. Flex-funding encourages the development of cooperative systems of care in rural areas, joining together CAHs, EMS providers, clinics, and health practitioners to increase efficiencies and quality of care.

Eligible Activities: The Flex program requires States to develop rural health plans and funds their efforts to implement community-level outreach and technical assistance to advance the following goals:

- improve quality of care and performance management;
- improve and integrate EMS;
- develop and implement rural health networks;
- support existing CAHs and eligible hospitals; and
- designate CAHs in the State.

Eligibility: Only States with CAHs or potential CAHs are eligible for the Flex program.

Website: www.hrsa.gov

Contact: HRSA, Office of Rural Health Policy
5600 Fishers Lane, Room 9A-42
Rockville, MD 20857
(301) 443-7322
Fax: (301) 443-2803

U.S. Department of Transportation (DOT)

Opportunity Title: Hazardous Materials Emergency Preparedness (HMEP) Grant

Purpose: The Hazardous Materials Emergency Preparedness (HMEP) grant program is intended to provide financial and technical assistance as well as national direction and guidance to enhance State, territorial, tribal, and local hazardous materials emergency planning and training. The HMEP grant program distributes fees collected from shippers and carriers of hazardous materials to emergency responders for hazmat training and to Local Emergency Planning Committees (LEPCs) for hazmat planning.

Eligible Activities: Hazardous materials emergency planning and training.

Eligibility: Grant funds are passed through to tribal and local emergency management offices having functional and active LEPCs and Transportation Environmental Resource Councils (TERCs).

Website: www.phmsa.dot.gov/hazmat/grants

Contact: U.S. Department of Transportation
Pipeline and Hazardous Materials Safety Administration
Hazmat Grants Program/PHH-50
East Bldg, 2nd Floor
1200 New Jersey Avenue Southeast
Washington, DC 20590
(202) 366-1109

Notes: Grant may require non-Federal matching funds.

U.S. Forest Service

Opportunity Title: Federal Excess Personal Property (FEPP) Program

Purpose: The FEPP program was enacted by Congress under the Federal Property and Administrative Services Act of 1949 and the Cooperative Forestry Assistance Act of 1978. The act directs the Secretary of Agriculture to encourage the use of FEPP to assist in reducing State fire budgets by loaning Federal-owned property to State foresters and their cooperators. To qualify for this program, a fire department must meet a number of requirements, one being a responsibility for providing fire protection on wildlands within the department's jurisdiction. The U.S. Forest Service monitors the use of loaned property to ensure that all FEPP acquired is used 90 percent for fire protection support and only 10 percent for other support.

Eligible Activities: Wildfire equipment.

Eligibility: Fire departments. Unlike the Volunteer Fire Assistance (VFA) Program, which benefits communities with a population at or below 10,000 residents, recipients of FEPP need only have wildland or rural fire responsibility that satisfies the State Forester.

Website: www.fs.fed.us/fire/partners/fepp/index.html

Contact: Contact your local State Forester for information.

U.S. Fire Administration (USFA)

Opportunity Title: Harvard Fire Executive Fellowship Program

Purpose: The Harvard Fire Executive Fellowship Program is sponsored through a partnership between the International Association of Fire Chiefs (IAFC), the International Fire Service Training Association (IFSTA), the National Fire Protection Association (NFPA), and the Department of Homeland Security—U.S. Fire Administration (DHS/USFA). Each organization funds the tuition costs for several scholarships.

Eligible Activities: Several senior fire executives will be awarded fellowships to attend Harvard's annual Program for Senior Executives in State and Local Government.

Eligibility: Senior fire executives.

Website: www.usfa.fema.gov/training/nfa/higher_ed/degree_programs/harvard

Contact: International Association of Fire Chiefs
ATTN: Harvard Fire Executive Fellowship Program
4025 Fair Ridge Drive, Suite 300
Fairfax, VA 22033-2868
(800) 238-3358, Extension 1072 (or 301)

U.S. Department of Housing and Urban Development (HUD)

Opportunity Title: HUD Good Neighbor Next Door Program

Purpose: Law enforcement officers, pre-Kindergarten through 12th-grade teachers, and firefighters/emergency medical technicians (EMTs) can contribute to community revitalization while becoming homeowners through HUD's Good Neighbor Next Door Sales Program. HUD offers a substantial incentive in the form of a discount of 50 percent from the list price of the home. In return, you must commit to live in the property for 36 months as your sole residence.

Eligible Activities: Purchase of eligible single-family homes located in revitalization areas. Full-time public safety officers may purchase a HUD-owned home at 50 percent of the market value of the home.

Eligibility: Full-time public safety officers including police officers, firefighters, and EMTs.

Website: www.hud.gov

Contact: U.S. Department of Housing and Urban Development
451 7th Street Southwest
Washington, DC 20410
(202) 708-1112

U.S. Department of Homeland Security (DHS)

Opportunity Title 1: Assistance to Firefighters Grants (AFG)

Purpose: AFGs provide financial assistance directly to fire departments and nonaffiliated EMS organizations to enhance their capabilities with respect to fire and fire-related hazards. Its primary goal is to help fire departments and nonaffiliated EMS organizations meet their firefighting and emergency-response needs. AFG seeks to support organizations that lack the tools and resources necessary to more effectively protect the life and safety of the public and their emergency-response personnel with respect to fire and all other hazards.

Eligible Activities: Fire department priorities include training, equipment, personal protective equipment (PPE) gear, firefighter wellness and fitness, modifications to fire stations and facilities, and firefighter vehicle acquisition. Nonaffiliated EMS organization priorities include EMS operations and safety, EMS training, EMS equipment acquisition, EMS PPE, EMS wellness and fitness, modifications to EMS stations and facilities, and EMS vehicle acquisition.

Eligibility: Fire departments and EMS organizations.

Website: www.fema.gov/firegrants

Contact: Fire department personnel who have questions regarding the AFGs can reach the Federal Emergency Management Agency's (FEMA's) Grant Programs Directorate (GPD) AFG Program staff at firegrants@dhs.gov

Notes: AFG has specific, population-based non-Federal matching requirements. See AFG guidelines for more information.

Opportunity Title 2: Staffing for Adequate Fire and Emergency Response (SAFER) Grants

Purpose: SAFER grants provide financial assistance to help fire departments increase their cadre of frontline firefighters. The goal is to assist local fire departments with staffing and deployment capabilities so they may respond to emergencies whenever they occur, assuring their communities have adequate protection from fire and fire-related hazards.

Eligible Activities: SAFER offers grants to support activities in two categories: 1) hiring of firefighters, and 2) recruitment and retention of volunteer firefighters.

Eligibility: Volunteer and combination fire departments, career fire departments, municipalities and fire districts, Statewide or local volunteer firefighter interests organizations.

Website: www.fema.gov/firegrants

Contact: Fire department personnel who have questions regarding the SAFER grants can reach FEMA's GPD AFG Program staff at firegrants@dhs.gov

Opportunity Title 3: Fire Prevention and Safety (FP&S) Grants

Purpose: The purpose of the AFG is to enhance the safety of the public and firefighters with respect to fire and fire-related hazards. The GPD also administers the FP&S grants as part of the AFG Program.

Eligible Activities: FP&S offers grants to support activities in two categories: 1) activities designed to reach high-risk target groups and mitigate incidences of deaths and injuries caused by fire and fire-related hazards (the "Fire Prevention and Safety Activity," and 2) research and development activities aimed at improvements of firefighter safety (the "Firefighter Safety Research and Development Activity").

Eligibility: Eligible applicants for this activity include fire departments, and national, regional, State, local, Native American tribal organizations, and/or community organizations that are recognized for their experience and expertise in fire prevention and safety programs and activities. Both private and public nonprofit organizations are eligible to apply for funding in this activity. For-profit organizations are not eligible to receive a FP&S grant award.

Website: www.fema.gov/firegrants

Contact: Fire department personnel who have questions regarding the FP&S grants can reach FEMA's GPD AFG Program staff at firegrants@dhs.gov

Opportunity Title 4: Pre-Disaster Mitigation (PDM) Program

Purpose: The PDM program provides funds to States, territories, Indian tribal governments, communities, and universities for hazard-mitigation planning and the implementation of mitigation projects prior to a disaster event. Funding these plans and projects reduces overall risks to the population and structures, while also reducing reliance on funding from actual disaster declarations. PDM grants are to be awarded on a competitive basis and without reference to State allocations, quotas, or other formula-based allocation of funds.

Eligible Activities: Predisaster planning and related activities.

Eligibility: States, U.S. territories, Indian and tribal governments, cities and townships, and universities.

Website: www.fema.gov/government/grant/pdm/index.shtm

Contact: FEMA Grant Program Directorate
Call Center
(866) 927-5646

Notes: The PDM program is subject to the availability of appropriation funding, as well as any directive or restriction made with respect to such funds. PDM grants are awarded to each State and territory as directed by Congress. **Cost-share:** up to 75-percent Federal cost-share. Small and impoverished communities may be eligible for up to a 90-percent Federal cost-share. Projects or initiatives that are eligible for funding under this announcement may or may not involve Geospatial issues (GIS).

Opportunity Title 5: Emergency Management Performance Grants (EMPG)

Purpose: The purpose of the EMPG is to assist State and local governments in enhancing and sustaining all-hazards emergency-management capabilities. An all-hazards approach to emergency response, including the development of a comprehensive program of planning, training, and exercises, provides the foundation for an effective and consistent response to any threatened or actual disaster or emergency, regardless of the cause.

Eligible Activities: Planning, training, exercises, and related all-hazards preparedness projects.

Eligibility: All 56 States and territories, as well as the Republic of the Marshall Islands and the Federated States of Micronesia.

Website: www.fema.gov/government/grant/empg/index.shtm

Contact: FEMA Grant Program Directorate
Call Center
(866) 927-5646

Notes: This grant has a non-Federal cost-sharing requirement.

Opportunity Title 6: Interoperable Emergency Communications Grant Program (IECGP)

Purpose: IECGP provides governance, planning, training, and exercise funding to States, territories, and local and tribal governments to carry out initiatives to improve interoperable emergency communications, including communications in collective response to natural disasters, acts of terrorism, and other manmade disasters. If a State Administrative Agent (SAA) and Statewide Interoperability Coordinator (SWIC)/Statewide Communication Interoperability Plan (SCIP) point of contact certify that its State or territory has fulfilled such governance, planning, training, and exercise objectives, the program provides the flexibility to purchase interoperable communications equipment with any remaining IECGP funds.

Eligible Activities: Planning, training, and exercises.

Eligibility: All 56 States and territories are eligible to apply for IECGP funds.

Website: www.fema.gov/government/grant/iecgp/index.shtm

Contact: FEMA Grant Program Directorate
Call Center
(866) 927-5646

Opportunity Title 7: Emergency Operations Center (EOC) Grant Program

Purpose: The EOC Grant Program is intended to improve emergency management and preparedness capabilities by supporting flexible, sustainable, secure, and interoperable EOCs with a focus on addressing identified deficiencies and needs.

Eligible Activities: This program provides funding for construction or renovation of a State, local, or tribal governments' principal EOC.

Eligibility: The governor of each State and territory is required to designate an SAA to apply for and administer the funds awarded under the EOC Grant Program. The SAA is the only eligible entity able to apply for the available funding on behalf of eligible State, local, and tribal EOCs.

Website: www.fema.gov/government/grant/eoc/index.shtm

Contact: FEMA Grant Program Directorate
Call Center
(866) 927-5646

Opportunity Title 8: Regional Catastrophic Preparedness Grant Program (RCPGP)

Purpose: The purpose of RCPGP is to enhance catastrophic incident preparedness in selected high-risk, high-consequence urban areas and their surrounding regions.

Eligible Activities: RCPGP is intended to support coordination of regional all-hazard planning for catastrophic events, including the development of integrated planning communities, plans, protocols, and procedures to manage a catastrophic event. The deliverables from the RCPGP will be made available throughout the country to enhance national resilience.

Eligibility: Eligible applicants for RCPGP include predesignated high-risk, high-consequence urban areas. The governor of each State and territory is required to designate an SAA to apply for and administer the funds awarded under RCPGP. The SAA is the only entity eligible to apply to FEMA for RCPGP funds.

Website: www.fema.gov/government/grant/rcp/index.shtm

Contact: FEMA Grant Program Directorate
Call Center
(866) 927-5646

Notes: Tier I RCPGP Sites include Bay area, Boston area, Chicago area, Houston area, Los Angeles/Long Beach, National Capital Region, New York City area, and Jersey City/Newark area. Tier II Sites include Honolulu area, Norfolk area, and Seattle area.

Opportunity Title 9: Tribal Homeland Security Grant Program (THSGP)

Purpose: The THSGP provides funds to directly eligible tribes to help strengthen the Nation against risks associated with potential terrorist attacks.

Eligible Activities: Activities related to preventing, preparing for, protecting against, and responding to acts of terrorism.

Eligibility: Any Native American community that operates a law enforcement or emergency services agency.

Website: www.fema.gov/government/grant/thsgp/index.shtm

Contact: FEMA Grant Program Directorate
Call Center
(866) 927-5646

Notes: Pursuant to the 9/11 Act, a directly eligible tribe applying for a grant shall designate an individual to serve as a tribal liaison with the DHS and other Federal, State, local, and regional government officials concerning preventing, preparing for, protecting against, and responding to acts of terrorism.

Opportunity Title 10: Homeland Defense Equipment Reuse (HDER) Program

Purpose: The HDER program provides responder agencies across the Nation access to a substantial inventory of radiological detection instrumentation and other equipment that is no longer required by the Federal government. This equipment is rehabilitated and provided at no cost to the recipient. In addition, training on the use of the equipment items and long-term support for their maintenance and field checks are available.

Eligible Activities: Radiological-detection equipment and related technical support and training.

Eligibility: Any State or local first-responder agency may place an order through their State or urban area HDER point of contact.

Contact: For more information about the HDER program or other initiatives of the Office of Grants and Training, contact the Centralized Scheduling and Information Desk (CSID) at (800) 368-6498 or askcsid@dhs.gov

Opportunity Title 11: Commercial Equipment Direct Assistance Program (CEDAP)

Purpose: The CEDAP complements the FEMA National Preparedness Directorate's other major grant programs to enhance regional-response capabilities, mutual aid, and interoperable communication by providing technology and equipment, together with training and technical assistance required to operate the equipment, to public safety agencies in smaller jurisdictions and certain metropolitan areas.

Eligible Activities: CEDAP funds equipment and training in five categories: extrication devices; thermal imaging, night vision, and video-surveillance tools; chemical-, biological-, and radiological-detection tools; information technology and risk management tools; vehicle-tracking tools.

Eligibility: Eligible applicants include law enforcement agencies and other emergency-responder agencies who demonstrate that the equipment will be used to improve their capability and capacity to respond to a major critical incident or to work with other first responders. Organizations must submit applications through the Responder Knowledge Base.

Website: www.ojp.sdoj.gov/odp

Contact: FEMA/CEDAP
Centralized Scheduling and Information Desk
245 Murray Lane, Building 410
Washington, DC 20528
(800) 368-6498

Homeland Security Grant Program (HSGP)

The HSGP suite consists of five subprograms, namely the State Homeland Security Program (SHSP), Urban Areas Security Initiative (UASI), Operation Stonegarden (OPSG), Metropolitan Medical Response System (MMRS), and Citizen Corps Program (CCP). These are different for the DHS grant programs discussed in the previous section of this report.

Opportunity Title 1: State Homeland Security Program (SHSP)

Purpose: This core assistance program provides funds to build capabilities at the State and local levels and to implement the goals and objectives included in State homeland security strategies and initiatives in their State Preparedness Report.

Eligible Activities: Consistent with the Implementing Recommendations of the 9/11 Act of 2007 (Public Law 110-53) (9/11 Act), States are required to ensure that at least 25 percent of SHSP appropriated funds are dedicated towards law enforcement terrorism prevention-oriented planning, organization, training, exercise, and equipment activities, including those activities which support the development and operation of fusion centers.

Eligibility: The SAA is the only entity eligible to apply to FEMA for SHSP funds. Recipients include all 50 States, the District of Columbia, Puerto Rico, American Samoa, Guam, Northern Mariana Islands, and the U.S. Virgin Islands.

Website: www.fema.gov/government/grant/hsgp/index.shtm

Contact: FEMA Grant Program Directorate
Call Center
(866) 927-5646

Opportunity Title 2: Urban Area Security Initiative (UASI)

Purpose: The UASI program focuses on enhancing regional preparedness in major metropolitan areas. The UASI program directly supports the national priority on expanding regional collaboration in the National Preparedness Guidelines and is intended to assist participating jurisdictions in developing integrated regional systems for prevention, protection, response, and recovery.

Eligible Activities: Consistent with the 9/11 Act, States are required to ensure that at least 25 percent of UASI-appropriated funds are dedicated towards law enforcement terrorism prevention-oriented planning, organization, training, exercise, and equipment activities, including those activities which support the development and operation of fusion centers.

Eligibility: The SAA is the only entity eligible to apply to FEMA for UASI funds. Prospective recipients for the UASI program include the 64 highest risk urban areas.

Website: www.fema.gov/government/grant/hsgp/index.shtm

Contact: FEMA Grant Program Directorate
Call Center
(866) 927-5646

Opportunity Title 3: Operation Stonegarden (OPSG)

Purpose: The intent of OPSG is to enhance cooperation and coordination among local, State, and Federal law enforcement agencies in a joint mission to secure the United States borders along routes of ingress from international borders to include travel corridors in States bordering Mexico and Canada, as well as States and territories with international water borders.

Eligible Activities: Border security.

Eligibility: Prospective recipients for OPSG include local units of government at the county level and Federally-recognized tribal governments in the States bordering Canada (including Alaska), southern States bordering Mexico, and States and territories with international water borders.

Website: www.fema.gov/government/grant/hsgp/index.shtm

Contact: FEMA Grant Program Directorate
Call Center
(866) 927-5646

Opportunity Title 4: Metropolitan Medical Response System Program (MMRS)

Purpose: The MMRS program supports the integration of emergency management, health, and medical systems into a coordinated response to mass-casualty incidents caused by any hazard. Successful MMRS grantees reduce the consequences of a mass-casualty incident during the initial period of a response by having augmented existing local operational response systems before the incident occurs.

Eligible Activities: EMS and emergency services preparedness and response projects for mass-casualty incidents.

Eligibility: The SAA is the only entity eligible to apply to FEMA for MMRS funds. Recipients for the MMRS program include the 124 MMRS jurisdictions.

Website: www.fema.gov/government/grant/hsgp/index.shtm

Contact: FEMA Grant Program Directorate
Call Center
(866) 927-5646

Opportunity Title 5: Citizen Corps Program (CCP)

Purpose: The Citizen Corps mission is to bring community and government leaders together to coordinate community involvement in emergency preparedness, planning, mitigation, response, and recovery.

Eligible Activities: Funding for CCPs. Refer to program guidance document for range of eligible activities.

Eligibility: The SAA is the only entity eligible to apply to FEMA for CCP funds. Recipients for the CCP include all 50 States, the District of Columbia, Puerto Rico, American Samoa, Guam, Northern Mariana Islands, and the U.S. Virgin Islands.

Website: www.fema.gov/government/grant/hsgp/index.shtm

Contact: FEMA Grant Program Directorate
Call Center
(866) 927-5646

Chapter 7: Foundations and Corporate Grants

There are a variety of private-funding sources available through nonprofit foundations and corporate-giving programs that may apply to emergency medical services (EMS) and fire services. Typically, these are one-time grants. Sometimes, the funding is multiyear. Virtually none are for continued support of general-operating expenses. However, private foundations and corporations are an excellent source of revenue providing grants for program planning, seed money for start-up costs, management and technical assistance, facility and equipment funding, and program-related investments. Private organizations are also a good source for in-kind donations of materials and services, and low-interest loans.

This chapter discusses major sources of funding from foundations and corporate-giving programs. It lists major grants available to improve fire and EMS in communities.

Foundations

There are tens of thousands of foundations in the United States. Foundations are nonprofit organizations that support charitable activities or community-based organizations that serve the common good. They are generally funded through endowments; money given by individuals, families, or corporations. Foundations are generally governed by a board of directors or trustees who oversee the financial assets and operations of their funds, and who decide how the foundation's money will be allocated.

Grants are funded from income earned from investments made with the endowment. The Internal Revenue Service (IRS) requires that independent and corporate foundations pay out at least 5 percent of the year-end fair market value of their assets. With rare exceptions, foundations may only donate money to tax-exempt 501(c)(3) organizations.

The Foundation Center (www.foundationcenter.org) is a leading source of information about foundations and philanthropic giving. Their website maintains a comprehensive database on United States and, increasingly, global grant-makers and their grants. The center operates research, education, and training programs to assist grant-seekers. In addition to the website, there are five regional libraries/learning centers and a network of 450 funding information centers located in public libraries, community foundations, and educational institutions nationwide. Fire chiefs and EMS managers interested in accessing private grants are highly encouraged to use this resource.

There are three basic types of grant-making foundations.

Independent Foundations

Independent foundations are the most common type of private foundation. They are not closely associated with a for-profit corporation. Independent foundations are formed from the philanthropy of an individual or family who establish an endowment to address a social or community cause. Smaller independent foundations may be operated by the donor or family members (family foundation); larger foundations typically have an independent board. The Foundation Center has identified 38,339 grant-making independent foundations. Often, independent foundations have a narrow focus such as funding the arts or education. Independent foundations that make grants available for community healthcare may be an excellent source of alternative funding for EMS agencies.

Corporate Foundations

Some for-profit companies funnel their charitable contributions and grants through a corporate foundation. Corporate foundations are separate legal entities, operated by an independent board of directors that is usually comprised of company executives and employees. Funding to corporate foundations can be made through an endowment, annual contributions from company profits, or a combination of both.

There are approximately 2,700 grant-making corporate foundations. These companies allocate billions of dollars annually for public service projects throughout the United States. National corporate foundations have the financial resources to provide generous grants, but the selection process is highly competitive. Their interests and application requirements are often well-developed and available to the public through annual reports, brochures, and websites.

While the size of some corporate foundations may be imposing, a well-developed plan may bring a large payoff for those determined enough to take on such a project. In formulating ideas, one should keep in mind the goals of the foundation and attempt to match organizational and community needs with those goals.

Community/Public Foundations

Community foundations are independent philanthropic institutions that serve a particular geographic area, typically a city or metropolitan region. Their mission tends to be more broadly defined than independent (or family) foundations seeking to improve the quality of life in their community. They receive their funding from a variety of individual donors and local businesses to establish endowment funds. Community/Public foundations are administered by a governing body or distribution committee representative of a variety of community interests.

Since community foundations are local, personal contact with board members can be very helpful when seeking grant funding for a project. Never underestimate the importance of personal contact.

Proposals for community foundation support should be based upon a well-defined local need. A narrowly-focused project is more likely to get funded by a small foundation than a broad project whose benefits are hard to evaluate.

When dealing with any foundation, independent (family), corporate, or community/public, be aggressive. Find out who the right people are and followup on all proposals with telephone calls or personal contact. EMS managers and fire chiefs should network within their community–in country clubs, social organizations, service clubs, and places of worship—to find out if any members of the community can help the department make contact with grant-makers.

Common Types of Foundation Grants

There are a variety of types of foundation grants. The following is a summary of the more common types.

General Operating Support Grants

Occasionally, a grant will be given to help support general-operating expenses. Organizations can use these grants to cover day-to-day activities or ongoing expenses such as administrative salaries, utilities, office supplies, etc. While general-operating support grants can be used at the discretion of the organization, grant-makers expect that all expenditures will be used to support the agency's overall mission.

Planning Grants

Planning grants are used to help organizations plan new programs. A planning grant can be used to fund initial project development work such as conducting a needs assessment, implementing a public participation program, acquiring consultant services, or conducting other planning activities.

Start-up Grants or Seed Money

Grants used to start new projects or organizations are called start-up grants or seed money. Nonprofit seed money can be used to cover administrative costs or other operating expenses of a new project. Start-up grants are awarded to new organizations or programs, typically in their first few years. Such grants can

be multiyear, with funding amounts decreasing each year. For example, a 3-year startup grant may award $30,000 the first year, $20,000 the second year, and $10,000 for the third year. After that, it is expected the organization or program will be self-sufficient or will have identified other sources of revenue.

Management Assistance Grants

Management assistance grants are meant to strengthen the management capacity of nonprofit organizations. Grants may allow organizations to retain technical assistance consultants who help in fundraising, marketing, financial management, etc. Departments might apply for a management assistance grant to help fund a fundraiser position.

Facilities and Equipment Grants

Facilities and equipment grants, sometimes referred to as capital-improvement grants, provide funding for "brick-and-mortar" projects such as building construction or equipment purchases such as ambulances and rescue vehicles.

Corporate-Giving

In addition to funding through corporate foundations, companies provide support through direct-giving programs. This can sometimes be a source of confusion to grant-seekers. Essentially, corporate-giving programs are not separate legal entities so they are not subject to the same IRS laws and foundations. They do not have an endowment; rather, they are funded through gifts from corporate profits or in-kind donations.

Corporations most commonly donate cash, but they may also donate the use of their facilities, property, services, or provide advertising support. They may also encourage employees to donate time or money.

Cash Donations

Corporations commonly donate cash. This money can be unrestricted and used for operational expenses or may come with stipulations for its use. Keep in mind that corporate-giving is usually closely aligned to the company's business interests or designed to benefit employees, their families, or the community where the company is located or does significant business.

Look for corporations with headquarters, regional offices, or branch locations within the jurisdiction. Some industries to consider include fast food restaurants, gasoline companies, factories, and insurance agents. Approaching a corporation is similar to a community foundation; keep an emphasis on personal contact. Being involved in the local Chamber of Commerce or community-based service organizations such as the Rotary Club, can put grant-seekers in contact with local company executives. These businessmen and women can help put the organization in contact with national offices.

When approaching corporations, it is helpful to provide a description of the potential benefits to the corporation, i.e., positive publicity, service to the community where the company does business, local use of company products or services, etc.

Many companies have programs to match donations made by their employees. Find out whether large firms which operate in the jurisdiction have such a program. If they do, target their employees for contributions and explain to them that their personal donations will be matched by their employer. In this way, local employees will feel they are supporting a local service and that the matching contribution will have an even larger effect.

In-Kind Contributions

Donations of goods and services are extremely valuable, especially for rural and volunteer organizations. Private companies of all sizes are often willing to donate in-kind services to support public safety projects, particularly if a deduction can be obtained. An in-kind donation of services might include a media production company creating a public service announcement (PSA) for radio or television. A local Webmaster might put together a webpage for the fire/EMS department. The local repair shop may donate mechanics, preventive-maintenance services, repairs, and parts.

Whether receiving cash or in-kind donations, it is important for the department to have a way to document and track donations. By law, nonprofit organizations cannot provide a donor with the dollar value of an in-kind donation. Such valuations are relative to the "fair market value" of the item and may need to be professionally assessed and certified elsewhere.

Executive Loan Programs

Another way a corporation can contribute to the local EMS or fire agency is to provide temporary executive talent. These programs make it possible for rural and volunteer departments to keep their overhead costs low, while providing an opportunity for business executives to invest in their community and build new executive leadership skills. In addition, the company enjoys increased visibility and goodwill in the community. Loaded executives come from all areas of the workforce, including middle management, clerical, and even retirees. Executive talent to access includes

- Chief executive officer;
- Chief financial officer;
- Chief operating officer;
- Corporate controller;
- Sales and marketing manager; and
- Business development manager.

Program-Related Investments

In addition to grants, cash, and in-kind donations, foundations may make loans, called program-related investments, to nonprofits. IRS rules govern program-related investments. Similar to grants, program-related investments can be used to provide funds to organizations that address community needs in line with the foundation's tax-exempt purpose and goals. Loans are usually made at low interest, or even zero interest. Unlike grants, program-related investments must be paid back to the foundation. Typically, program-related investments are made for capital-improvement projects such as building projects.

Philanthropy and Private Grants

The following is a list of philanthropic and private-grant sources that may be able to assist EMS and fire service agencies.

Opportunity Title 1: Automated External Defibrillation (AED) Grant

Organization: www.AEDGrant.com

Purpose: The AED grant is a corporate buy-down grant program funded by generous donations, corporate backing, and AED manufacturer sponsors. This program is designed to help institutions and individuals purchase AED devices at affordable prices.

Eligible Activities: AED purchases.

Eligibility: Private and public businesses, agencies, and institutions. Individuals are also welcomed to apply.

Website: www.aedgrant.com

Contact: AEDGrant.com
565 Westlake Street, Building 100
Encinitas, CA 92024
(760) 944-1048

Opportunity Title 2: Allstate Foundation

Organization: Allstate Foundation.

Purpose: This foundation invests millions of dollars annually in the United States. The Allstate Foundation is dedicated to fostering safe and vital communities where people live, work, and raise families. These communities are economically strong, crime-free, and residents feel a sense of belonging and commitment.

Eligible Activities: Grant priorities fall into three categories: Safe and Vital Communities, Economic Empowerment and Tolerance, Inclusion and Diversity. A majority of Allstate's funding is dedicated to two signature issues: Teen Safe Driving and Domestic Violence.

Eligibility: 501(c)(3) organizations; Federal and State organizations; municipalities.

Website: www.allstate.com/foundation.aspx

Contact: The Allstate Foundation
2775 Sanders Road, F4
Northbrook, IL 60062

Notes: The Allstate Foundation has different applications for 501(c)(3) organizations and municipalities.

Opportunity Title 3: Bedford Community Health Foundation Grant

Organization: Bedford Community Health Foundation.

Purpose: This foundation serves the health-related needs of the citizens of Bedford City, VA, and surrounding communities.

Eligible Activities: Rescue squad support, emergency care, EMS equipment.

Eligibility: Nonprofit health-care organizations in the Bedford, VA, area.

Website: www.healthybedford.org

Contact: Bedford Community Health Foundation
P.O. Box 1104
Bedford, VA 24523
(540) 586-5292

Notes: Grant is limited to applicants from the Bedford, VA area.

Opportunity Title 4: Brant Chesney Memorial Grant Program

Organization: Georgia Firefighters Burn Foundation (GFBF).

Purpose: The GFBF has established a special grant program in memory of Gwinnett County career firefighter, Brant Chesney, who lost his life on December 27, 1996, battling a fire during his off-duty hours as a volunteer with the Forsyth County Fire Department. Grants awarded are to be used by fire departments in the State of Georgia to develop a new program or enhance an existing public life safety program, dedicated primarily to fire safety and burn prevention education.

Eligible Activities: Fire prevention and life safety education programs.

Eligibility: Applicants must be 1) a Georgia-certified fire department, and 2) have project goal(s) achievable within a 6-month period after funding is awarded.

Website: www.gfbf.org

Contact: Georgia Firefighters Burn Foundation
2575 Chantilly Drive
Atlanta, GA 30324
(404) 320-6223
Fax: (404) 320-6190

Notes: Grant is limited to fire departments in Georgia.

Opportunity Title 5: Building Healthy Communities Grant Program

Organization: The Home Depot Foundation.

Purpose: Grants, up to $2,500, are now available to registered 501(c)(3) nonprofit organizations, public schools, or tax-exempt public service agencies in the United States who are using the power of volunteers to improve the physical health of their community.

Eligible Activities: Grants are made in the form of The Home Depot gift cards for the purchase of tools or materials.

Eligibility: Only 501(c)(3) nonprofit organizations, public schools, or tax-exempt public service agencies in the United States are eligible to apply.

Website: http://corporate.homedepot.com/wps/portal/Grants

Contact: Email: community_impact@homedepot.com

Notes: Grants must support work completed by community volunteers. Work must be completed within 6 months of the grant award.

Opportunity Title 6: Contributing to Our Communities Grant

Organization:	Farmers' Insurance Group.
Purpose:	Farmers' invites proposals from programs that improve safety, expand educational opportunities, enhance health and human services, encourage civic participation, and support arts and culture in States where Farmers' does business.
Eligible Activities:	Open grant process.
Eligibility:	Preference is given to 501(c)(3) organizations.
Website:	www.farmers.com/What_we_believe.html
Contact:	Community Relations Manager Farmers' Insurance Group 4680 Wilshire Boulevard Los Angeles, CA 90010

Opportunity Title 7: Disaster Grants

Organization:	Do Something Organization.
Purpose:	This program provides funding directly for youth projects. The community-action grants help turn dream projects into a reality and take existing projects to the next level. Past grantees have used the money to create a community-run organic farm, publish a youth-written literary magazine for women of color, and even create an organization that teaches sick kids how to fly.
Eligible Activities:	Projects focused on disaster preparedness, emergency response, rebuilding after a disaster, and supporting U.S. military troops.
Eligibility:	Individuals 25 or under. Applicants must be U.S. or Canadian citizens.
Website:	www.dosomething.org/grants/disaster
Contact:	Do Something Organization (212) 254-2390, Extension 238

Opportunity Title 8: E-One Tell Your Story Fire Truck Grant

Organization:	Emergency One, Inc.
Purpose:	The E-One Tell Your Story Fire Truck Grant is a competition that gives fire departments the opportunity to win a new commercial pumper by demonstrating a need.
Eligible Activities:	Fire apparatus competition.
Eligibility:	Fire departments.
Website:	www.e-one.com/news/eone-stories/tell-your-story.htm

Contact: E-One, Inc.
1601 Southwest 37th Avenue
Ocala, FL 34474
(352) 237-1122
Fax: (352) 237-1151

Opportunity Title 9: Firehouse Subs Equipment Grant

Organization: Firehouse Subs Public Safety Foundation.

Purpose: The Firehouse Subs Public Safety Foundation is dedicated to improving life safety capabilities of emergency-service entities in communities served by Firehouse Subs by providing funding, resources, and support to public safety entities.

Eligible Activities: Equipment.

Eligibility: Public safety departments.

Website: www.firehousesubs.com/content.cfm?id=388

Contact: Firehouse Subs Public Safety Foundation
Attn: Grant Request
3400 Kori Road
Jacksonville, FL 32257
(904) 886-8300

Notes: Grants are limited to communities served by Firehouse Subs.

Opportunity Title 10: Firefighters Charitable Foundation Grants

Organization: Firefighters Charitable Foundation.

Purpose: The Firefighters Charitable Foundation continues to provide assistance to those in need. Grants are given to assist local fire/disaster victims, fire prevention education, volunteer fire department equipment purchase, and community safety programs.

Eligible Activities: Grants are available for the following programs: AEDs, community smoke-detector programs, fire department equipment, and juvenile firesetter programs.

Eligibility: Fire departments.

Website: www.ffcf.org/grants.html

Opportunity Title 11: Fireman's Fund Heritage Program

Organization: Fireman's Fund Insurance Company.

Purpose: The Fireman's Fund Heritage is a national community-based program that provides funds for equipment, fire prevention tools, firefighter training, fire safety education, and community emergency-response programs. Fireman's Fund employees and agents award grants and provide volunteer support for local fire departments, national firefighter organizations, and burn prevention/treatment organizations. Created by Fireman's Fund Insurance Company as its philanthropic mission, the Fireman's Fund Heritage program supports firefighters for safer communities.

Eligible Activities: Fire equipment, fire prevention, firefighter training, fire safety education, and community emergency-response programs.

Eligibility: Fire departments.

Website: www.firemansfund.com

Contact: Fireman's Fund Insurance Company
(866) 440-8716

Opportunity Title 12: Fire Safety Pledge Program

Organization: Liberty Mutual Insurance.

Purpose: The Fire Safety Pledge program is an interactive fire safety resource for families from Liberty Mutual. Ten $10,000 Fire Safety Pledge awards will be awarded to the fire departments in the communities that show the most support.

Eligible Activities: Nonspecific.

Eligibility: Individuals complete an online Fire Safety Pledge quiz and credit their completed quiz to give a point to their local fire department. The fire departments with the most points in their respective division will earn a grant. Grants are available to any fire department in the 50 States, including the District of Columbia (excluding Guam and Puerto Rico).

Website: www.befiresmart.com/Protect-Your-Community/Fire-Safety-Pledge.aspx

Contact: Email: team@befiresmart.com

Notes: Grants will be awarded based on community population.

Opportunity Title 13: FM Global Fire Prevention Grant Program

Organization: FM Global.

Purpose: FM Global provides financial support to organizations working to combat fire. Through the Fire Prevention Grant Program, fire departments and fire brigades, as well as national, State, regional, local, and community organizations can apply for funding to support a wide array of fire prevention, preparedness, and control efforts.

Eligible Activities: Prefire planning for commercial, industrial, and institutional facilities; fire and arson prevention and investigations; fire prevention education and training programs.

Eligibility: Fire departments and brigades, as well as national, State, regional, local, and community organizations.

Website: www.fmglobal.com/page.aspx?id=01060200

Contact: Email: firepreventiongrants@fmglobal.com

Opportunity Title 14: Georgia-Pacific Bucket Brigade Grant

Organization: Georgia-Pacific.

Purpose: The Georgia-Pacific Bucket Brigade Grant supports volunteer and smalltown fire departments through monetary grants for equipment, resources, and programming, as well as safety education materials used in schools.

Eligible Activities: Grants have been used to fund a variety of necessities, including turnout gear, extrication equipment, vehicle repairs or purchases, water pumps, hoses and nozzles, programming initiatives, and more.

Eligibility: Communities with Georgia-Pacific facilities or communities where Georgia-Pacific employees reside.

Website: www.gpbucketbrigade.com

Contact: Email: gpbucketbrigade@gapac.com

Opportunity Title 15: Helen Close Charitable Foundation

Organization: Helen Close Charitable Foundation.

Purpose: Giving primarily for scholarships, a ranch experience for inner-city children, police/public safety, and animal interests.

Eligible Activities: Disasters, fire prevention and control, and equipment.

Eligibility: Public safety organizations.

Contact: Helen Close Charitable Foundation
100 West Liberty Street, Suite 1100
Reno, NV 89501-1959
(775) 786-6141

Notes: Eligible only to northern Nevada and adjacent counties in northern California.

Organization Title 16: International Fire Relief Mission (IFRM)

Purpose: The IFRM is a 501(c)(3) nonprofit corporation whose mission is to provide lifesaving assistance to active-duty firefighters and rescue personnel in countries lacking adequate education or equipment; provide specialized demonstrations on using donated fire and rescue equipment in accordance with internationally-recognized standards; and ensure successful, long-term assistance by developing international partnerships and collaborations for continued support, preparation, and accountability at the receiving country.

Eligible Activities: Fire and EMS equipment including personal protective equipment (PPE), self-contained breathing apparatus (SCBA), compressors, rescue tools, apparatus, and EMS equipment.

Eligibility: Fire and public safety organizations outside the United States.

Website: www.ifrm2007.com/

Contact: International Fire Relief Mission
30345 Elm Street
Lindstrom, NM 55045
(612) 669-8500

Opportunity Title 17: The Jeremiah Lucey Grant Program

Organization: The Leary Firefighters Foundation.

Purpose: The Jeremiah Lucey Grant Program, named in honor of actor Denis Leary's cousin who perished in a catastrophic fire in Worcester, MA, provides financial support for the training and equipment needs of uniformed firefighters in board-designated geographic areas.

Eligible Activities: Training and equipment purchases.

Eligibility: Only union-affiliated, uniformed fire departments in board-designated geographic areas are eligible to apply.

Website: www.learyfirefighters.org

Contact: The Leary Firefighters Foundation
594 Broadway, Suite 409
New York, NY 10012
(212) 343-0240

Opportunity Title 18: Lacy and Connor Search and Rescue Fund

Organization: Carole Sund/Carrington Memorial Reward Foundation.

Purpose: The purpose of the fund is to provide financial assistance to law enforcement agencies and nonprofit organizations involved in search-and-rescue operations. These agencies are encouraged to apply for financial assistance for training, equipment, or any item that can improve their search-and-rescue efforts.

Eligible Activities: Equipment and training, including PPE, field equipment, medical-care equipment, rescue equipment, and dive equipment.

Eligibility: Organizations involved in search-and-rescue operations.

Website: www.carolesundfoundation.com/index.php

Contact: The Carole Sund/Carrington Memorial Fund Foundation
301 Downey Avenue
Modesto, CA 95354
(209) 567-1059

Opportunity Title 19: Motorola Foundation Public Safety and Security Grant Program

Organization: The Motorola Solutions Foundation.

Purpose: The Motorola Solutions Foundation provides grants to support public safety workers, communities, and families in the United States and Canada. Examples of programs funded include support for families of first responders killed or injured in the line of duty, first-responder training, crisis hotlines, and financial support for volunteer fire departments.

Eligible Activities: Disaster preparedness, disaster relief, fire safety, scholarships for fallen firefighters, and law enforcement officers' dependents.

Eligibility: Eligible organizations may include local chapters of police and fire foundations, community disaster preparedness, or fire safety programs and scholarship programs for fallen law enforcement officers or firefighters and others.

Website: http://responsibility.motorolasolutions.com/

Contact: Motorola Solutions Foundation
1303 East Algonquin Road
Schaumburg, IL 60196
Phone: (847) 576-6200
Fax: (847) 576-9440
Email: foundation@motorolasolutions.com

Notes: Grants cannot be used to purchase Motorola equipment or services.

Opportunity Title 20: Nationwide Insurance Grant Program

Organization: Nationwide Insurance Foundation.

Purpose: The Nationwide Insurance Foundation is dedicated to making strategic philanthropic investments to meet the most critical needs in communities.

Eligible Activities: The foundation makes funding decisions based on a Community Impact Model. In order of priority: Tier 1: emergency and basic needs; Tier 2: crisis stabilization; Tier 3: personal and family empowerment; and Tier 4: community enrichment. Grant types include general-operating support, project and/or program support, and capital support.

Eligibility: The Nationwide Insurance Foundation considers funding organizations in a limited number of cities. Refer to the grant website for details.

Website: www.nationwide.com/about-us/what-we-fund.jsp

Opportunity Title 21: State Farm Safe Neighbors Grant

Organization: State Farm Insurance.

Purpose: State Farm Insurance provides company grants in three focus areas: safety, community development, and education. In addition to these three focus areas, there is very limited funding to meet community-based needs including, but not limited to, environment, health and wellness, and other safety issues.

Eligible Activities: State Farm Safety Grants are directed toward automobile and roadway safety, home safety and fire prevention, disaster preparedness, disaster recovery, and personal financial safety and security.

Eligibility: Grants are awarded to nonprofit, tax-exempt organizations under Section 501(c)(3) of the U.S. Internal Revenue Code, Canadian charitable organizations, educational institutions, and governmental entities.

Website: www.statefarm.com/aboutus/community/grants/company/company.asp

Opportunity Title 22: U.S. Smokeless Tobacco Company (USSTC) Operational Ranger Program

Organization: USSTC.

Purpose: This program awards rugged, off-road utility vehicles to the Nation's emergency responders.

Eligible Activities: Acquisition of off-road vehicles.

Eligibility: Applying organizations must be based in the United States. The following organizations are eligible for consideration as part of this donation program: volunteer and career fire departments, police and Sheriff departments, wildlife conservation and enforcement, EMS and ambulance organizations, search and rescue, and park or forest rangers.

Website: www.ussmokeless.com/en/cms/Responsibility/Investing_in_Communities/default.aspx

Opportunity Title 23: Verizon Foundation Grant

Organization: Verizon Foundation.

Purpose: The Verizon Foundation targets our philanthropic investments to partners and programs that achieve meaningful outcomes and measurable results in specific areas within each of our funding priorities.

Eligible Activities: Education, literacy, domestic violence prevention, health care and accessibility, and Internet safety.

Eligibility: Applicants must be a 501(c)(3) organization or public school. Applications may be accepted from some government agencies.

Website: http://foundation.verizon.com/grant/guidelines.shtml

Contact: Questions about the program should be directed to your local Verizon Community Relations Manager.

Opportunity Title 24: Walgreen Community Grant Program

Organization: Walgreens Pharmacies.

Purpose: The Walgreen Community Grant Program provides funding for programs that focus on access to community health and wellness programs, pharmacy education programs and mentoring initiatives, civic and community outreach initiatives, and emergency and disaster relief.

Eligible Activities: Community health and wellness programs; emergency and disaster relief.

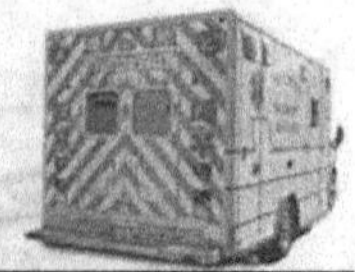

Eligibility: Limited to nonprofit, tax-exempt organizations under Section 501(c)(3) of the Internal Revenue Code.

Website: www.walgreens.com/marketing/about/community/guidelines.jsp

Contact: Email: grants@walgreens.com

Opportunity Title 25: The Provident Bank Foundation Grant

Organization: The Provident Bank Foundation.

Purpose: The Provident Bank Foundation supports organizations involved with arts and culture, education, health, recreation, and human services.

Eligible Activities: Health and wellness—The foundation supports programs designed to provide essential, underlying medical care; be nonduplicative of similar programs in the region; and focus on physical and mental wellness for people in all age categories.

Social and Civic Services—The foundation supports programs designed to improve the quality of life in the community and support the underlying social infrastructure, including volunteer organizations that provide services vital to the community such as fire departments, rescue squads, and health-care agencies.

The foundation has provided funding for building construction and renovation, capital campaigns, equipment purchases, general-operating support, management development, program development, scholarship funds, and project-seed money.

Website: www.providentnjfoundation.org/index.aspx

Contact: The Provident Bank Foundation
830 Bergen Avenue
Jersey City, NJ 07306
(201) 915-5434

Appendix A: Successful Grant Application Example

Enhancing Porter Township Emergency Services

Porter Township Emergency Services
P.O. Box 409
Union, MI 49130
616-641-7100 fax616-641-2571

Amount of funds requested $11,453
Federal ID: 38-2441694
Submitted: 04-20-2000
For Period: July 1, 2000 to Sept. 30, 2001

Contact: Michael W. Holdeman
P.O. Box 409
Union, MI 49130
616-641-7100

TABLE OF CONTENTS

RURAL HEALTH INITIATIVE (RHI) FUNDING
ABBREVIATED EQUIPMENT AND EDUCATIONAL ASSISTANCE
GRANT APPLICATION
Applications due: April 21, 2000

The Abbreviated Equipment and Educational Assistance Grant application can be used by those eligible organizations who are interested in making equipment, supplies, materials or education/training requests. **Project requests for RHI funds must be for $25,000 or less, and provide the specified matching funds required by subsection A, B, C, or D in Section III - Statement of Work.**

1. **Project Title:** Enhancing Porter Township Emergency Services

2. **My organization is applying for funds from the following area(s):** *(check all that apply)*

☒ Rural EMS

☐ Rural Community Collaboration

☐ Rural Network Development & Telecommunications

☐ Rural Non-emergency Transportation

3. **Amount of Request:**
Please enter in the amount of funds requested, the match contribution and the budget for each of the funding areas:

	Rural EMS	**Rural Community Collaboration**	**Rural Network Development & Telecommunications**	**Rural Non-Emergency Transportation**
Amount Requested	$11,453			
Match Contribution	$11,453			
Budget	$22,906			

4. **Total Budget:**
Enter in the project's total amount requested from all funding pools, total match contribution provided and project's total budget

Total Grant Amount Requested $11,453 *(NOT to exceed $25,000)*
+
Total Match Provided $11,453
=
Total Project Budget $22,906 *(Grant Request plus Match)*

1

5. **Name of Applicant Organization:** Porter Township Emergency Services

Authorized Official: Larry Carroll

Title: Supervisor, Porter Township

Mailing Address: P.O. Box 517

City/County: Union Cass State Michigan Zip 49130

Telephone: (616) 641-2375 Fax (616) 641-2571

Email Address: porter@portertownship.org

6. **Contact Person:** Michael W. Holdeman

Title: Fire Chief, Porter Township Emergency Services

Mailing Address: P.O. Box 409

City/County: Union Cass State Michigan Zip 49130

Telephone: (616) 641-7100 Fax (616) 641-2571

Email Address:

7. **Legal Status of Organization** *(check only one response)*

Private, Non-Profit Entity (attach copy of IRS's 501(C)(3) or other legal documentation verifying status)

x Public Entity (attach copy of IRS letter verifying Federal Tax ID number)

8. **Federal Tax ID Number:** 38-2441694

9. **Authorizing Entity** – I hereby affirm my authority and responsibility for the use of all equipment and/or educational training described in this application.

Larry L. Carroll

Authorized Individual *(signature)*

Larry L. Carroll, Supervisor

Printed Name

April 20, 2000

Date

2

10 Needs Statement (15 points) - Describe the problem(s) to be addressed. What factors contribute to the problem(s)? Include relevant demographic or health related data that substantiates that the problem(s) exists. What barriers have prevented the problem(s) from being successfully resolved? Why are RHI funds needed? Attach a mep of the service area.

Porter Township Emergency Services is a Governmental entity wholly owned and operated by Porter Township, Cass County, Michigan. It is comprised of Porter Twp. Fire Department, and Porter Twp. EMS, currently providing Fire and EMS services to the residents of South Porter and Mason Townships. The service area spans approximately 65 square miles (see map) of mostly residential, rural, and Lake Residential, with a very small commercial, industrial base. Serving a growing population of approximately 5,500 over the entire service area, several lakes complicate the process of delivering service due to difficult geographic locations and difficult to navigate roads. This proposal defines a plan to address some of the issues within our control to improve and enhance our delivery of services.

Problem #1 We are plagued as most volunteer organizations are with a lack of well-trained EMT, and MFR volunteers. The decline in Volunteerism noted nation wide hits home to our mostly rural communities, adding to this the lack of and lack of proximity to training facilities is discouraging to potential volunteers considering the commitment of time for the necessary training required to deliver quality care. While many steps have already been taken by our organization, including the hiring of 1 part-time and 1 full-time EMT to assist in covering day shifts, more needs to be done to encourage the commitment of qualified individuals within the community.

Problem #2 Geographic Location, affecting the timely delivery of critical services. Our numerous lakes create some difficulty in rapid delivery of some very time critical emergency services, such as BLS, trauma care, and early CPR, early defibrillation. The extremely time critical nature of such services prompts us to explore alternatives to our current service delivery system.

While we are fairly successful with very early warning cardiac related incidents via BLS transport and ALS intercepts, actual defibrillation attempts have been relatively unsuccessful with 2 successful defibrillation cases out of 22 attempts. We feel this is partly due to the time required to assemble the EMS crew and travel to the scene.

11 Project Description/Workplan (20 Points) - Describe the project and how it will impact the problem(s) identified in the needs statement. Describe how RHI funds will be used and the direct benefit that will result. List other benefits that may also result. Provide a detailed list of items to be purchased on the *Items List* sheet that follows. What is the expected life of those items?

To address problem 1, we have initiated the on site review process with SWM Systems, Regional Coordinator, Paul Dickens. Plans are being implemented to complete the necessary steps to approve our facility to host MFR, and EMT-Basic classes in house. This will make it considerably more convenient for potential volunteers to commit to attending the necessary education, and continuing education required without the travel/time issues we now face. Additionally CPR classes will be scheduled for the winter of 2000, open to the public and hosted at our facility to increase the level of community preparedness, as well as building awareness and interest in EMS.

RHI funds are needed to purchase the needed equipment as listed for hosting EMT, MFR as well as CPR classes once the site review process has been completed, and the site/classes have been approved.

RHI funds will also be used to cover tuition for community CPR classes and MFR/EMT classes, minimizing volunteers out of pocket expenses.

In addressing Problem 2 we have considered many issues, including personnel as addressed above, specifically more trained individuals within the community, as well as more residents with CPR training. This should effectively address the first "link" in the cardiac incident survivability namely early CPR.

Secondly placement of the proper equipment in trained hands within the EMS system. RHI funds will provide MFR and EMT's with trauma, and oxygen kits, as well as hand held portable radios, in order to provide them with the equipment needed to respond quickly directly to an incident while the on duty EMS crew is assembling.

Thirdly, the placement of 2 additional AED's. One AED purchased with RHI, plus matching funds from EMS department capitol equipment funds will be placed on an MFR vehicle. This vehicle will be available as first out vehicle for a single MFR or EMT to respond independent of the on duty EMS crew. This also gives us a vehicle capable of responding to calls when the Ambulance may be on another call, thus delivering early patient care during possible prolonged response time while waiting for another service. This MFR vehicle is already in the possession of the Fire Department, and is currently being readied for service and equipped with the necessary medical equipment, with guidelines set by SWM Systems. Much of the necessary equipment has already been purchased with budgeted operating funds; the remainder of the necessary equipment items are addressed in the equipment purchases proposed by this project.

The second AED would be issued to equip a trained EMT who resides in a remote location of our service area. These

steps will address the second link of cardiac survivability, namely early defibrillation.

These steps coupled with our support of the proposed CCALS system will better serve our residents by preparing us for rapid delivery of critical services when needed. Even more specifically would be the cardiac emergency readiness, continued training and early CPR, early Defibrillation, and early ALS delivers the best possible care to the residents of our communities.

RHI funds will also be used to purchase equipment to supply a proposed 8 additional MFR's and EMT's with full field response kits.

Life expectancy of all items proposed is at least 10 years, with the exception of consumables included in field response kits.

List items to be purchased in order of priority on the following table. Provide as much information about each piece as possible. If applicable, include: make, model, etc. Supplies, materials and equipment prices should be based upon three informal bid requests (via phone, fax, etc.) with the lowest price listed.

ITEM LIST

(Listing brands will not preclude applicant from having to obtain bids for generic equipment types.)

Quantity	Description	Local Share	State Share
2	Physio control lifepack 500 AED	$3,950	$3,950
1	Noonin 5800 pulse oximeter	$350	$350
2	(1 sager, 1 hare) Traction Splints	$300	$300
8	Field kits complete (o2 & trauma)	$2000	$2000
3.	Motorola HT1000 hand held radios	$1200	$1200
1	Portable Suction Device	$265	$265
1	Set Airway Trainers	$890	$890
1	Set CPR Trainers	$250	$250
2	(long boards, short boards, straps & imob. devices) Sets Immobilization Devices	$298	$298
1	AED Trainer for above AED's	$175	$175
1	Stair Chair	$275	$275
	(see attached page for tuition expenses)		
	Total	$9,953	* $9,953

*NOT to exceed $25,000

Total Project Budget $22,906

List items to be purchased in order of priority on the following table. Provide as much information about each piece as possible. If applicable, include: make, model, etc. Supplies, materials and equipment prices should be based upon three (3) informal bid requests (via phone, fax, etc.) with the lowest price listed.

ITEM LIST

(Listing brands will not preclude applicant from having to obtain bids for generic equipment types.)

Quantity	Description	Local Share	State Share
	Special Attachment Page for Tuition Costs		
4	EMT Students @$450	$900	$900
6	MFR Students @$100	$300	$300
50	CPR Students (community members)	$300	$300
	XXXXX Total Tuition Totals	$1,500	* $1,500

*NOT to exceed $25,000

Total Project Budget $22,906

Work Plan – State the overall goal of the project, and list objectives, activities and the quarter in which the activities or purchases are expected to be completed in the appropriate column.

Project Goal: More rapid delivery of Critical EMS Services

MFR Vehicle in service MFR's Trained and in service	Order additional AED's Train in use of new AED's Order additional portable radios/field kits Purchase field resp. kits, deploy equip/vehicle	1st Qtr.
Additional AED's	Additional AED's in service/purchase	2nd Qtr.
On Site Review	On site review process completed. site course approval Training equipment purchased. (aed trainer, traction splints, boards...)	3rd Qtr
Community offered CPR Classes	Community open recert, and initial classes	3rd Qtr.
EMT Class	First on-site EMT class undervay	4th Qtr.

*Quarters:

1s quarter = July 1 – Sept 30, 2000
2nd quarter = Oct. 1 - Dec. 31, 2000
3rd quarter = Jan. 1 – March 31, 2001
4th quarter = April 1 – June 30, 2001
5th quarter = July 1 - Sept. 30, 2001

MICHIGAN DEPARTMENT OF COMMUNITY HEALTH
PROGRAM BUDGET - COST DETAIL

DCH-0386 6/98
Replaces FIN-116

PAGE 1 OF 1

Program	Code	Budget Period	Date Prepared
Rural Health Initiative		Jul. 1/00 to Sep.30 01	04-19-00

Contractor	Original Budget	Amended Budget	Amendment Number
Porter Township Emergency Services	XX		

POSITION DESCRIPTION	POSITIONS REQUIRED	TOTAL SALARY	COMMENTS
TOTAL	0	$0.00	

FRINGE BENEFITS:
(Specify) FICA ____ UNEMPLY INS. ____ RETIREMENT ____ HOSP. INS. ____
LIFE INS. ____ VISION INS. ____ HEARING INS. ____ OTHER. ____
DENTAL INS. ____ WORK. COMP. ____
COMPOSITE RATE
AMOUNT ____

TRAVEL (Specify if any item exceeds 10% of Total Expenditures) ____

SUPPLIES & MATERIALS (Specify if any items exceeds 10% of Total Expenditures) ____

CONTRACTUAL (Subcontracts) TOTAL SUB-CONT ____

EQUIPMENT (Specify) See equipment detail attached pg. 6
TOTAL EQUIPMENT $19,906

OTHER EXPENSES (Specify if any item exceeds 10% of Total expenditures:
Communications
Space Cost
Other Tuitions, see attached pg.7 TOTAL OTHER $3,000

OTHER COST DISTRIBUTIONS			INDIRECT COST CALCULATIONS
Description of cost being distributed:	Percent distributed to the program:	Amount Distributed	
	____%	$____	BASE X RATE
	____%	$____	
	____%	$____	BASE X RATE

12

12 Collaboration (15 point) - What organizations are involved in the project? Describe the various roles and responsibilities that each of the partners will assume. How will project efforts be coordinated and communicated to other organizations? How has the community been involved and how will they be informed about the projects progress? Attach a list of project partners.

This project is focused on the organization of Porter Township Emergency Services, as this organization will carry the majority of the weight for implementing the project. It is not however without collaboration. Other organizations involved with the program are Porter Township Community Policing Committee, and Mason Township Community Policing Committee. These community-based committees will be instrumental in support of fund raising efforts, as well as direct communication to residents via newsletters and various meetings and community events sponsored and participated in by their organizations. SWM Systems will be conducting the on site review, and coordinating the instructor/coordinator recruitment, as well as approving education programs, and licensing vehicles and individuals. Cass County Medical Control is involved in reviewing the performance and with setting protocols for responding units/individuals in the county. CCALS, Bristol FD. and Osolo EMS (all ALS providers) will continue to provide ALS intercept service.

13 Project sustainability/Match (20 points) - Describe how the project will be sustained once RHI funds end. Briefly explain how maintenance and other related expenses will be sustained over the life of the item(s). Applications should also describe how pertinent training issues will be addressed. Describe what type(s) of match have been included and their source. Also indicate the amount of the match and to which funding pool it corresponds.

RHI funding of this project is mostly limited to the purchase of equipment, most of which holds a life expectancy of 10 years or more. Maintenance and operating expenses of this equipment over the expected 10-year life will be appropriated from the operating budget of the Porter Township EMS annual budget. All department equipment is inspected on a weekly, monthly and annual basis and maintained accordingly.

Porter Township Emergency Services hosts regular re-certification classes as part of its regular training schedule, and additional funding is budgeted annually for training.

Matching funds are sourced from budgeted capital equipment funds, and community fund raising/donations.

14 Outcomes Measures and Evaluation (15 points) - Describe what major outcomes are expected as a result of the project. How will outcomes be monitored and reported? What types of data will be used? How will the data be collected and how often?

Our main goal for this project is to save more lives in our community. That being a difficult goal to measure statistically given the innumerable variables with each individual given case, the only way to measurably reach this type of goal is to control as many of those variables as possible. The funding of this grant will allow us to directly affect 3 of those variables. 1-Training, 2-Response times, and 3-Available Equipment.

The first- Training; Measured by enrollment and finally the roster of licensed trained responding members of our EMS service. Data is maintained in a central database developed for the fire and EMS industry. This data will be analyzed and reported to the necessary officials quarterly, and at the end of the grant period.

Second- Response time; Data is recorded and analyzed in the same fashion as training, using the same database. AED information is downloaded after every use into a computer database and will be analyzed for goal criteria as well. Each AED will be downloaded into the same database to provide for continuity of records.

The database is currently in place and in use, and proven to be an effective tool in analyzing response data. Detailed information is available upon request.

And third- Available Equipment; is well addressed in the other two categories of data collection. For example, extended response times are sometimes attributed to the lack of available AED.

15 Budget Narrative (15 points) - Describe the amount of funds requested and any cost sharing among partners. Be sure to include all revenues necessary to support the proposed projects. Briefly explain how funds requested are aligned with the project's goals and objectives.

Porter Township Emergency Services is requesting $10,453 in RHI funds from the Rural Emergency Medical Services pool. Fund expenditures will be as follows.

To support the work plan as described in the Project Description/Work Plan the following, proposed expenditures and funding is as follows, more detail on equipment is attached on the Item list supplied.

	Local Funds	RHI Funds	Total
Equipment Total	$9,953	$9,953	$19,906
Tuitions (Includes 4 EMT's at $450 6 MFR's at $100 50 CPR at $12)	$1,500	$1,500	$ 3,000
TOTALS	$11,453	$11,453	$22,906

CPR class materials include Manuals, Exams, and CPR masks for each student.
The total proposed budget for this project is $22,906. 50% or $11,453 being funded by RHI funds, and 50% or $11,453 being funded by budget departmental funds. These funds will allow us to address the work-plan described, and effectively meet the projects objective of timely delivery of critical services.

MICHIGAN DEPARTMENT OF COMMUNITY HEALTH
PROGRAM BUDGET SUMMARY

PAGE 1 OF 1

Program	Code	Budget Period	Date Prepared
Rural Health Initiave		Jul.1 00 TO Sep.30 2001	04-19-00
Contractor: Porter Twp. Emergency Services		Budget for Original Agreement XXX or Amendment # ____	
Address: P.O. Box 409	City: Union State: Michigan	Zip Code: 49130	Payee Identification Number: 38-2441694

	EXPENDITURE CATEGORY	RHI Reral EMS	Local Matching		TOTAL BUDGET
1	Salaries & Wages				
2	Fringe Benefits				
3	Travel				
4	Supplies & Materials				
5	Contractual (Subcontracts)				
6	Equipment	$9,953	$9,953		$19,906
7	Other Expenses: Tuitions	$1,500	$1,500		$3,000
8	TOTAL DIRECT	$11,453	$11,453		$22,906 $0.00
9	Indirect Costs: Rate #1				
	Indirect Costs: Rate #2				
10	Other Cost Distributions				
11	TOTAL EXPENDITURES	$11,453	$11,453		$22,906 $0.00
	SOURCE OF FUNDS				
12	Fees & Collections				
13	State Agreement	$11,453			
14	Local		$11,453		
15	Federal				
16	Other				
17	TOTAL FUNDING	$11,453	$11,453		$22,906 $0.00

Completion is a Condition of Funding

13

Porter Township

To Whom It May Concern:

This is to certify that Porter Township, Cass County, Michigan is a governmental unit and therefore, not subject to taxes.

Marty Russell

Marty Russell
Porter Township Clerk
P. O. Box 517
Union, MI 49130

#38-2441697

Attachment A

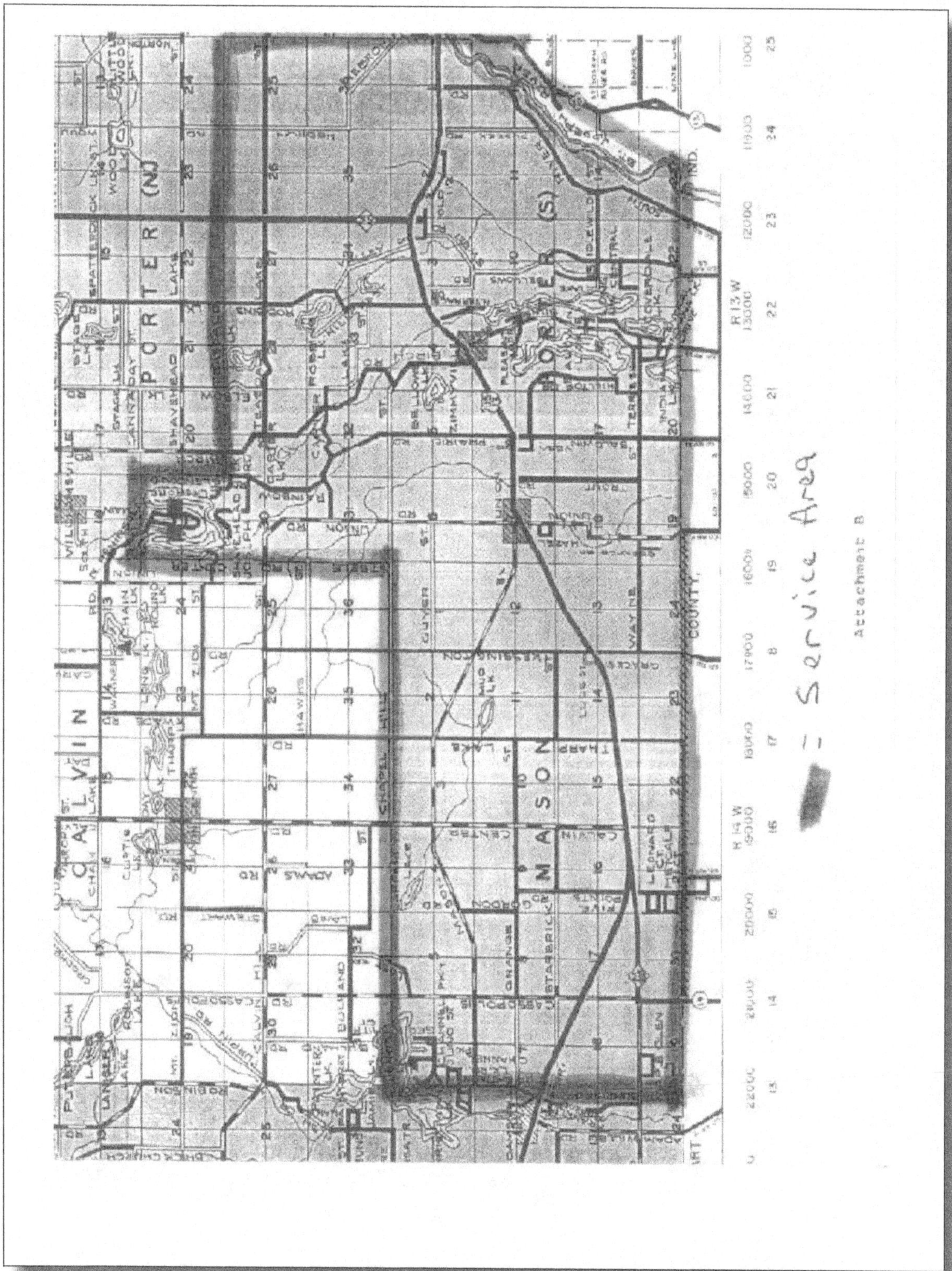

▬ = Service Area

Attachment B

OFFICE OF SHERIFF

Cass County, Michigan

JOSEPH M. UNDERWOOD JR.
Sheriff

LARRY GORHAM
Undersheriff

April 19, 2000

Michigan Department of Community Health
Community Assessment Section
Health Legislation and Policy Development
6th Floor, Lewis Cass Building
320 S. Walnut Street
Lansing, MI 48913

Dear Ms. Christner

Porter Township Emergency Services is the primary provider of EMS services in the communities of Porter and Mason Townships, providing twenty-four hour seven day per week Fire, Rescue and EMS services.

Porter Township Emergency Services has demonstrated their desire to cooperate with our organization and others in the area in providing these necessary services as well as enhance other health related services in their community.

We continue to support the efforts of the Porter Township Emergency Services in increasing services in our community.

Should you have any questions or concerns, please feel free to contact me. Thank you for your time on this important issue.

Sincerely,

Joseph M. Underwood, Jr
Sheriff

JMU/dd

321 M-62 North • Cassopolis, MI 49031 • Telephone (616) 445-8644
Fax Numbers (616): Sheriff/445-0036; Administration/445-1254; Dispatch/445-2484; Booking/445-0882

Attachment C

Cass County Community Policing • Porter Division
V.P.T.S.C.P. • A Not For Profit Corporation
P.O. Box 506 • Union, MI 49130-0506

April 17, 2000

Michigan Department of Community Health,

Community Assessment Section,
Health Legislation and Policy Development
6th Floor, Lewis Cass Building
320 S. Walnut Street
Lansing, MI 48913

Ms. Theresa Christner:

Porter Community Policing Committee is a non-profit community based committee developed to further public safety and community based policing in Porter Township.

Porter Township Emergency Services is the primary provider of EMS services in our community, providing 24 hour 7-day a week Fire, Rescue, and EMS services.

Porter Township Emergency Services has demonstrated their desire to cooperate with our organization and others in the area in providing these necessary services as well as enhance other health related services in our community.

We continue to support the efforts of the Porter Township Emergency Services in increasing services in our community.

Should you have any questions or concerns please feel free to contact me. Thank you for your time on this important matter.

Sincerely,

Coe Strain
Chairperson
Porter Township CPC
P.O. Box 506
Union, MI 49130
(616) 641-5931

Attachment D

Appendix B: State Emergency Medical Services Offices

Alabama
EMS Division
AL Department of Public Health
The RSA Tower, 201 Monroe Street, Suite 750
Montgomery, AL 36130-3017
(334) 206-5383
www.alapubhealth.org/

Alaska
Section of Common Health and EMS
410 Willoughby, Suite 109
Juneau, AK 99811-0616
(907) 465-3027
www.chems.alaska.gov/

Arizona
Arizona Department of Health, Bureau of EMS
150 North 18th Avenue, Suite 540
Phoenix, AZ 85007
(800) 200-8523
(602) 364-3150
www.azdhs.gov/bems/

Arkansas
Department of Health, EMS and Trauma
4815 West Markham Street, Slot 38
Little Rock, AR 72205-3867
(501) 661-2178
www.healthy.arkansas.gov

California
California EMS Authority
1930 9th Street
Sacramento, CA 95814
(916) 322-4336
www.emsa.ca.gov/

Colorado
Colorado Department of Health & Environment–EMSDP-PCP-A-5
4300 Cherry Creek Drive
Denver, CO 80246-1530
(303) 692-2991
www.cdphe.state.co.us/em/emhom.html

Connecticut
Connecticut Office of EMS
410 Capitol Avenue
P.O. Box 340308
Hartford, CT 06134-0308
(860) 509-7975
www.ct.gov/dph/site/default.asp

Delaware
Delaware Office of EMS, BHCC
655 Bay Road, Suite 4-H
Dover, DE 19901
(302) 739-4710
www.dhss.delaware.gov/dph/ems/

Florida
Florida Department of Health, Bureau of EMS
2002-D Old Saint Augustine Road
Tallahassee, FL 32301-4881
(850) 245-4444
www.doh.state.fl.us/

Georgia
Georgia Office of EMS
2 Peachtree Street, Suite 12-432
Atlanta, GA 30303
(404) 679-0547
http://ems.ga.gov/

Hawaii
State Emergency Medical Services and Injury Prevention System Branch
3675 Kilauea Avenue
Honolulu, HI 96816
(808) 733-9210
http://hawaii.gov/health/family-child-health/ems/index.html

Idaho
Idaho EMS Bureau
590 West Washington
Boise, ID 83702
(208) 334-4000
www.healthandwelfare.idaho.gov/

Illinois
Division of EMS & Highway Safety,
Department of Public Health
525 West Jefferson Street
Springfield, IL 62761
(217) 785-2080
www.idph.state.il.us/

Indiana
Indiana Department of Homeland Security
EMS Certifications, E239, IGC-S
302 West Washington Street
Indianapolis, IN 46204-2739
(800) 666-7784
www.in.gov/dhs/3525.htm

Iowa
Iowa Department of Public Health–Bureau of EMS
401 Southwest 7th, Suite D
Des Moines, IA 50309
(515) 725-0322
www.idph.state.ia.us/ems/

Kansas
Kansas Board of EMS
Landon State Office Building
900 Southwest Jackson, Room 1031
Topeka, KS 66612
(785) 296-7296
www.accesskansas.org/ssrv-ksbemsRenew/index.do

Kentucky
Kentucky Board of Emergency Medical Services
275 East Main Street, HS1E-F
Frankfort, KY 40621-0001
(502) 564-8963
www.kbems.org/index.htm

Louisiana
Bureau of EMS
Louisiana Department of Health & Hospitals
P.O. Box 94215
Baton Rouge, LA 70804
(225) 763-5700
www.dhh.louisiana.gov/offices/?id=220

Maine
Maine Board of Emergency–Medical Services
Central Maine Commerce Center
45 Commerce Drive, Suite 1
152 State House Station
Augusta, ME 04333-0152
Jan Brinkman, RN, EMT-P
Education & Training Coordinator
www.state.me.us/dps/ems/

Maryland
Maryland Institute for Emergency
Medical Services Systems
653 West Pratt Street
Baltimore, MD 21201
(800) 762-7157
www.rmiemss.org/home/

Massachusetts
Massachusetts Department of Public Health
Office of Emergency Medical Services
2 Boylston Street, 3rd Floor
Boston, MA 02116
(617) 753-7300
www.mass.gov/dph/oems

Michigan
Licensing Division
MDCIS, Bureau of Health Services
201 Townsend Street–Capitol View Building
P.O. Box 30717
Lansing, MI 48913
(517) 335-0918
www.michigan.gov/mdch/0,1607,7-132-2946_5093_28508---,00.html

Minnesota
Minnesota EMS Regulatory Board
2829 University Avenue, Southeast, #310
Minneapolis, MN 55414-3222
(612) 627-6000
www.emsrb.state.mn.us/

Mississippi
Mississippi State Department of Health,
Emergency Medical Services
P.O. Box 1700
Jackson, MS 39215-1700
(601) 576-7366
http://msdh.ms.gov/msdhsite/_static/47.html

Missouri
Missouri Department of Health & Senior Services,
Bureau of EMS
P.O. Box 590
Jefferson City, MO 65102
(573) 751-6356
http://health.mo.gov/safety/ems/index.php

Montana
Montana Department of Public Health &
Human Services
EMS & Trauma Services Section
1400 Broadway Street, Cogswell Building
Room C-317
Helena, MT 59620
(406) 444-3895
www.dphhs.mt.gov/ems/

Nebraska
NEDDHS, EMS Division
301 Centennial Mall, South
Lincoln, NE 68509-5007
(402) 471-3578
www.hhs.state.ne.us/ems/emsindex.htm

Nevada
Department of Health & Human Resources, EMS
850 Elm Street
Elko, NV 89801
(775) 753-1154
http://health.nv.gov/EMS_Contacts.htm

New Hampshire
Department of Safety, Div. of EMS
Northern New Hampshire Field Office
55 Maynesboro Street
Berlin, NH 03570
(603) 752-7531
www.nh.gov/safety/divisions/fstems/

New Jersey
Department of Health & Senior Services, OEMS
50 East State Street
Trenton, NJ 08625-0360
(609) 633-7777
www.state.nj.us/health/ems/index.shtml

New Mexico
New Mexico Emergency Medical Systems
1301 Silver Road, Building F
Santa Fe, NM 87507
(505) 476-8200
www.nmems.org/

New York
New York State Department of Health,
Bureau of EMS
Corning Tower
Empire State Plaza
Albany, NY 12237
(518) 402-0996
www.health.state.ny.us/nysdoh/ems/main.htm

North Carolina
Office of EMS
2707 Mail Service Center
Raleigh, NC 27699-2707
(919) 855-3935
www.ncems.org/

North Dakota
Division of Emergency Health Services,
North Dakota State Health Department
600 East Boulevard Avenue, Department 301
Bismarck, ND 58505-0200
(701) 328-2388
www.ndhealth.gov/ems/

Ohio
Ohio Division of EMS
P.O. Box 182073
1970 West Broad Street
Columbus, OH 43218-2073
(614) 466-9477
www.ems.ohio.gov/

Oklahoma
State Department of Health, EMS Division
1000 Northeast 10th Street
Oklahoma City, OK 73117-1299
(405) 271-4027
www.ok.gov/health/Protective_Health/Emergency_Medical_Services/

Oregon
Oregon Health Division, EMS & Trauma Systems
P.O. Box 14450
Portland, OR 97214-0450
(503) 731-4011
www.ok.gov/health/Protective_Health/Emergency_Medical_Services/

Pennsylvania
Department of Health, EMS Office
P.O. Box 90
Harrisburg, PA 17108
(717) 787-8740
www.pa-ems.org/

Rhode Island
Rhode Island Department of Health, EMS
3 Capitol Hill, Room 105
Providence, RI 02908-5097
(401) 222-2401
www.health.state.ri.us/

South Carolina
Division of EMS, SD DEC
2600 Bull Street
Columbia, SC 29201
(803) 545-4204
www.scdhec.gov/health/ems/

South Dakota
Office of EMS
600 East Capitol
Pierre, SD 57501-2536
(605) 773-4031
http://doh.sd.gov/

Tennessee
Department of Health, Division of EMS
426 5th Avenue North, 1st Floor Cordell Hull Building
Nashville, TN 37247-0701
(615) 741-2584
http://health.state.tn.us/ems/

Texas
Texas Department of State Health Services
1100 West 49th Street
Austin, TX 78756-3199
(512) 834-6700
www.dshs.state.tx.us/emstraumasystems.default.shtm

Utah
Department of Health, Bureau of EMS
288 North, 1460 West 2nd Floor
Salt Lake City, UT 84114-2852
(801) 538-6287
www.health.state.ut.us/ems/

Vermont
Vermont Department of Health, EMS
P.O. Box 70
Burlington, VT 05402
(802) 863-7310
http://healthvermont.gov/

Virginia
Office of Emergency Medical Services
Virginia Department of Health
1041 Technology Park Drive
Glen Allen, VA 23059
www.vdh.virginia.gov/oems/

Washington
Emergency Medical Services & Trauma Systems
P.O. Box 47853
Olympia, WA 98504-7853
(360) 236-2800
www.doh.wa.gov/haqa/emstrauma/

West Virginia
Office of EMS
350 Capitol Street, Room 515
Charleston, WV 25301-3716
(304) 558-3210
www.wvdhhr.org/ocrhs/

Wisconsin
Wisconsin EMS Section
1 West Wilson Street, Room 372
P.O. Box 2659
Madison, WI 53701-2659
(608) 266-0472
www.dhs.wisconsin.gov/ems/

Wyoming
Department of Health, EMS Office
Hathaway Building, 4th Floor
Cheyenne, WY 82002
(307) 777-7955
www.health.wyo.gov/sho/ems/index.html

American Samoa
Department of Public Health
LBJ Medical Center, American Samoa Government
Pago Pago, AS 96799
(011) (684) 633-5003

Northern Mariana Islands
Department of Public Safety - Fire Division
P.O. Box 10001
Saipan, MP 96950
(670) 664-9081

Puerto Rico
Emergency Medical System
P.O. Box 2702
San Juan, PR 00931-2702
(787) 766-1733

Virgin Islands
Emergency Medical Services/DOH
#48 Sugar Estate
Charlotte Amalie, VI 00802
(809) 776-7708, Extension 4706

Appendix C: State Fire Marshal Offices

Alabama
Department of Insurance
P.O. Box 303351
Montgomery, AL 36130-3351
(334) 241-4166
www.aldoi.org/FireMarshal/aspx

Alaska
Department of Public Safety
5700 East Tudor Road
Anchorage, AK 99507-1225
(907) 269-5491
www.dps.state.ak.us/Fire/

Arizona
Department of Building and Fire Safety
1110 West Washington, Suite 100
Phoenix, AZ 85007
(602) 364-1003
www.dfbls.az.gov/contacts.aspx

Arkansas
State Fire Marshal
#1 State Police Plaza Drive
Little Rock, AR 72209
(501) 618-8700
www.arfireprevention.org

California
Department of Forestry & Fire Protection
1131 "S" Street
Sacramento, CA 95814
(916) 445-8200
www.fire.ca.gov/fire_prevention/fire_prevention.php

Colorado
Division of Fire Safety
690 Kipling Street, Suite 2000
Denver, CO 80215
(303) 239-4600
http://dfs.state.co.us/Contact.htm

Connecticut
Christopher Guari, Acting Fire Marshal
Department of Public Safety
1111 Country Club Road
Middletown, CT 06457
www.ct.gov/dps/cwp/view.asp

Delaware
State Fire Marshal's Office
1537 Chestnut Grove Road
Dover, DE 19904-1544
(302) 739-5665
http://statefiremarshal.delaware.gov/

Florida
Office of the State Fire Marshal
200 East Gaines Street
Tallahassee, FL 32399-0340
(352) 369-2800
www.myfloridacfo.com/sfm/

Georgia
Department of Insurance and Fire Safety
2 Martin Luther King Jr. Drive, Suite 620, West Tower
Atlanta, GA 30334
(404) 656-2064
www.gainsurance.org/FireMarshal/Home.aspx

Hawaii
State Fire Council
636 South Street
Honolulu, HI 96813
(808) 723-7151
http://hawaii.gov/labor/sfc

Idaho
Department of Insurance
700 West State Street, 3rd Floor
P.O. Box 83720
Boise, ID 83720-0043
(208) 334-4250
www.doi.idaho.gov/sfm/firemars.aspx

Illinois
Office of the State Fire Marshal
1035 Stevenson Drive
Springfield, IL 62703
(217) 785-0969
www.state.il.us/osfm/

Indiana
Department of Homeland Security
402 West Washington Street, Room E208
Indianapolis, IN 46204-2243
(317) 232-2222
www.in.gov/dhs/

Iowa
State Fire Marshal's Office - DPS
215 East 7th Street, 3rd Floor
Des Moines, IA 20319
(515) 725-6145
www.dps.state.ia.us/fm/building/index.shtml

Kansas
State Fire Marshal
700 Southwest Jackson Suite 600
Topeka, KS 66603
(785) 296-3401
www.ksfm.ks.gov

Kentucky
Fire Prevention
101 Sea Hero Road, Suite 100
Frankfort, KY 40601
502-573-0382
http://dhbrc.ky.gov/fp/

Louisiana
Department of Public Safety
8181 Independence Boulevard
Baton Rouge, LA 70806
(800) 256-5452
http://sfm.dps.louisiana.gov

Maine
State Fire Marshal's Office
52 State House Station
Augusta, ME 04333-0052
(207) 626-3870
www.maine.gov/dps/fmo/

Maryland
Maryland State Police
1201 Reisterstown Road
Pikesville, MD 21208
(410) 653-8980
http://firemarshal.state.md.us/

Massachusetts
Department of Fire Services
P.O. Box 1025
Stow, MA 01775
(978) 567-3111
www.mass.gov/

Michigan
Bureau of Fire Service
300 North Washington Square, 4th Floor
Lansing, MI 48913
(517) 241-8847
www.michigan.gov/lara/0,1607,7-154-28077_42271_42321---,00.html

Minnesota
Department of Public Safety
444 Cedar Street, Suite 145
Saint Paul, MN 55101
(651) 201-7200
https://dps.mn.gov/divisions/sfm/Pages/default.aspx

Mississippi
Department of Insurance
P.O. Box 79
Jackson, MS 39205
(601) 359-1061
www.mid.state.ms.us/state_fire_marshal/state_fire_marshal_office.aspx

Missouri
Division of Fire Safety
P.O. Box 844
Jefferson City, MO 65102
(573) 751-2930
www.dfs.dps.mo.gov/DFSStaff.htm

Montana
Fire Prevention - Department of Justice
P.O. Box 201415
Helena, MT 59620-1415
(406) 444-2050
www.doj.mt.gov/enforcement/fireprevention/default.asp

Nebraska
State Fire Marshal
246 South 14th Street
Lincoln, NE 68508-1804
(402) 471-2027
www.sfm.state.ne.us/

Nevada
Office of the State Fire Marshal
107 Jacobsen Way
Carson City, NV 89711
(775) 684-7500
http://fire.state.nv.us/Administration.shtml

New Hampshire
Department of Public Safety
33 Hazen Drive
Concord, NH 03305
(603) 271-3294
www.nh.gov/safety/divisions/firesafety/contactus.html

New Jersey
Division of Fire Safety
P.O. Box 809
Trenton, NJ 08625-0809
(609) 633-6106
www.state.nj.us/dca/divisions/dfs/

New Mexico
Office of the State Fire Marshal
1120 Paseo De Peralta
Santa Fe, NM 87501
(505) 427-0066
www.nmprc.state.nm.us/sfm.htm

New York
NYS Office of Fire Prevention and Control
99 Washington Avenue, Suite 500
Albany, NY 12231-0001
(518) 474-6746
www.dhses.ny.gov/ofpc/

North Carolina
DOI, Office of the State Fire Marshal
1202 Mail Service Center
Raleigh, NC 27699-1202
(919) 661-5880
www.ncdoi.com/osfm/default.asp

North Dakota
State Fire Marshal's Office
P.O. Box 1054
Bismarck, ND 58502-1054
(701) 328-5555
www.ag.nd.gov/fm/FM.htm

Ohio
Department of Commerce
8895 East Main Street
Reynoldsburg, OH 43068-3395
(614) 752-8200
http://com.ohio.gov/fire/

Oklahoma
Office of the Oklahoma State Fire Marshal
2401 Northwest 23rd, Suite 4
Oklahoma City, OK 73107
(405) 522-5011
www.oklaosf.state.ok.us/~firemar/

Oregon
State Fire Marshal's Office
4760 Portland Road, Northeast
Salem, OR 97305
(503) 378-3473
www.oregon.gov/OSP/SFM/index.shtml

Pennsylvania
Office of the State Fire Commissioner
2605 Interstate Drive
Harrisburg, PA 17110
(717) 651-2201
www.osfc.state.pa.us/portal/server.pt/community/state_fire_commissioner_home/4462

Rhode Island
Office of the State Fire Marshal
118 Parade Street
Providence, RI 02909
(401) 462-4200
www.fire-marshal.ri.gov/

South Carolina
Division of Fire & Life Safety
141 Monticello Trail
Columbia, SC 29203
(803) 896-9800
www.scfiremarshal.llronline.com

South Dakota
Office of the State Fire Marshal
118 West Capitol Avenue
Pierre, SD 57501
(605) 773-3562
http://dps.sd.gov/emergency_services/state_fire_marshal/default.aspx

Tennessee
Division of Fire Prevention
500 James Robertson Parkway, 3rd Floor
Nashville, TN 37243
(615) 741-2981
www.state.tn.us/commerce/sfm

Texas
State Fire Marshal's Office
P.O. Box 149221
Austin, TX 78714-9221
(512) 305-7900
www.tdi.texas.gov/fire/index.html

Utah
Department of Public Safety, Office of the SFM
P.O. Box 141775
Salt Lake City, UT 84114-1775
(801) 965-4461
http://publicsafety.utah.gov/firemarshal/

Vermont
Division of Fire Safety
1311 US Route 302 - Berlin, Suite 600
Barre, VT 05641-2351
(802) 479-7561
www.dps.state.vt.us/fire

Virginia
Department of Housing and
Community Development
1005 Technology Park Drive
Glen Allen, VA 23059
(804) 371-0220
www.vafire.com/state_fire_marshal/index.html

Washington
Washington State Patrol, Fire Protection Bureau
P.O. Box 42600
Olympia, WA 98504-2600
(360) 596-4000
www.wsp.wa.gov/fire/firemars.htm

West Virginia
State Fire Marshal's Office
1207 Quarrier Street, 2nd Floor
Charleston, WV 25301
(304) 558-2191
www.wvfiremarshal.org

Wisconsin
Department of Justice
P.O. Box 7857
Madison, WI 53707-7857
(608) 266-1671
www.doj.state.wi.us/dci/arson

Wyoming
Department of Fire Prevention and Electrical Safety
Herschler 1 West
Cheyenne, WY 82002
(307) 777-7288
www.wyofire.state.wy.us

Guam
Guam Fire Department
1301-1 Central Avenue
P.O. Box 2950
Hagatna, Guam 96932
(671) 472-3304
www.gfd.guam.gov

Northern Mariana Islands
Department of Public Safety–Fire Division
P.O. Box 10001
Saipan, MP 96950
(670) 664-9081

Puerto Rico
Puerto Rico Fire Department
Cuerpo de Bomberos de Puerto Rico
Calle Concepcion #23
Guayanilla, PR 00656
(787) 835-2330

United States Virgin Islands
Virgin Island Fire Services
6002 East Diamond Ruby Suite 7
Christiansted, VI 00820
(240) 773-8050

Appendix D: State Homeland Security Contacts

Alabama
Alabama Department of Homeland Security
P.O. Box 304115
Montgomery, AL 36130-4115
(334) 956-7250
Main Fax: (334) 223-1120
www.dhs.alabama.gov

Alaska
Office of Homeland Security
Department of Military and Veteran Affairs
P.O. Box 5750
Building 4900, Suite B-214
Ft. Richardson, AK 99505
(907) 428-7062
www.ak-prepared.com/

Arizona
Arizona Department of Homeland Security
1700 West Washington Street, #210
Phoenix, AZ 85007
(602) 542-7030
www.azdohs.gov

Arkansas
Department of Emergency Management
Building 9501
Camp Joseph
North Little Rock, AR 72119-9600
(501) 683-6700
www.adem.arkansas.gov

California
California Emergency Management Agency
3650 Schriever Avenue
Mather, CA 95655
(916) 324-9809
www.oes.ca.gov/

Colorado
Colorado Department of Public Safety
700 Kipling Street
Denver, CO 80215
(303) 239-4398
www.cdpsweb.state.co.us

Connecticut
Department of Emergency Management and Homeland Security
25 Sigourney Street, 6th Floor
Hartford, CT 06106-5042
(860) 256-0800 or (800) 397-8876
www.ct.gov/demhs/

Delaware
Delaware Homeland Security Advisor
Department of Safety and Homeland Security
303 Transportation Circle
P.O. Box 818
Dover, DE 19903
(302) 744-2680
http://dshs.delaware.gov/

District of Columbia
Homeland Security &
Emergency Management Agency
2720 Martin Luther King Jr. Avenue Southeast
Washington, DC 20032
(202) 727-4036
www.dcema.dc.gov/dcema/site/default.asp

Florida
Florida Department of Law Enforcement
P.O. Box 1489
Tallahassee, FL 32302-1489
(850) 410-8300
www.fdle.state.fl.us/osi/DomesticSecurity/

Georgia
Georgia Emergency Management Agency/
Homeland Security
P.O. Box 18055
Atlanta, GA 30316-0055
(404) 635-7000
www.gema.state.ga.us

Hawaii
Hawaii State Civil Defense
3949 Diamond Head Road
Honolulu, HI 96816-4495
(808) 733-4246
www.scd.state.hi.us

Idaho
Bureau of Homeland Security
4040 West Guard Street, Building 600
Boise, ID 83705-5004
(208) 422-3040
www.bhs.idaho.gov

Illinois
Illinois Emergency Management Agency
2200 South Dirksen Parkway
Springfield, IL 62703
(217) 557-6225
http://iema.illinois.gov/iema/news/news.asp

Indiana
Indiana Department of Homeland Security
302 West Washington Street, Room E-208
Indianapolis, IN 46204
(317) 232-3986
www.in.gov/sema/

Iowa
Iowa Department of Homeland Security
7105 Northwest 70th Avenue
Camp Dodge Building W-4
Johnston, IA 50131
www.iowahomelandsecurity.org

Kansas
Kansas Homeland Security
2800 Southwest Topeka
Topeka, KS 66611-1287
(785) 274-1001
www.kansastag.gov/kshls_default.asp

Kentucky
Office of Homeland Security
200 Metro Street
Frankfort, KY 40622
(502) 564-2081
http://homelandsecurity.ky.gov/

Louisiana
Governor's Office of Homeland Security
and Emergency Preparedness
7667 Independence Boulevard
Baton Rouge, LA 70806
(225) 925-7500
www.ohsep.louisiana.gov

Maine
Department of Defense, Veterans &
Emergency Management
33 State House Station, Camp Keyes
Augusta, ME 04333-0001
(207) 626-4205
www.maine.gov/mema/homeland/

Maryland
Governor's Homeland Security Office
The Jeffrey Building
16 Francis Street
Annapolis, MD 21401
(800) 492-8477
www.gov.state.md.us/homelandsecurity.html

Massachusetts
Executive Office of Public Safety and Security
1 Ashburton Place, Room 2133
Boston, MA 02108
(617) 727-7775
www.mass.gov/?pageID=eopshomepage&L=1&L0=
Home&sid=Eeops

Michigan
Homeland Security Advisor
333 South Grand Avenue
P.O. Box 30634
Lansing, MI 48909
(517) 241-0401
www.michigan.gov/homeland/

Minnesota
Homeland Security & Emergency Management
444 Cedar Street
St. Paul, MN 55101
(651) 201-7400
www.hsem.state.mn.us/

Mississippi
Office of Homeland Security
P.O. Box 958
Jackson, MS 39296-4501
(601) 346-1499
www.homelandsecurity.ms.gov

Missouri
Missouri Office of Homeland Security
P.O. Box 749
Jefferson City, MO 65102
(573) 522-3007
www.dps.mo.gov/homelandsecurity/

Montana
Montana Disaster & Emergency Services Division
1956 Mount Majo Street
P.O. Box 4789
Ft. Harrison, MT 59636
(406) 841-3911
http://dma.mt.gov

Nebraska
Nebraska Emergency Management Agency
1300 Military Road
Lincoln, NE 68508-1090
(402) 471-7421
www.nema.ne.gov/index.shtml

Nevada
Nevada Homeland Security
2478 Fairview Drive
Carson City, NV 89711
(775) 687-0300
http://homelandsecurity.nv.gov/

New Hampshire
Division of Homeland Security
and Emergency Management
33 Hazen Drive
Concord, NH 03305
(800) 735-2964
www.nh.gov/safety/divisions/hsem/index.html

New Jersey
New Jersey Homeland Security and Preparedness
P.O. Box 091
Trenton, NJ 08625
(609) 584-4000
www.njhomelandsecurity.gov

New Mexico
New Mexico Department of Homland Security &
Emergency Management
13 Bataan Boulevard
Santa Fe, NM 87504
or
P.O. Box 27111
Santa Fe, NM 87502
(505) 476-1051
http://nmdhsem.org/

New York
New York State Division of Homeland Security
1220 Washington Avenue
State Office Campus
Building 7A Suite 710
Albany, NY 12242
(518) 242-5000
www.dhses.ny.gov/oct/

North Carolina
Department of Crime Control and Public Safety
4701 Mail Service Center
Raleigh, NC 27699
(919) 733-2126
www.nccrimecontrol.org

North Dakota
Division of Homeland Security
Department of Emergency Services
P.O. Box 5511
Bismarck, ND 58506
(701) 328-8100
www.nd.gov/des/

Ohio
Homeland Security
1970 West Broad Street
Columbus, OH 43223-1102
(614) 387-6171
www.homelandsecurity.ohio.gov

Oklahoma
Oklahoma Office of Homeland Security
P.O. Box 11415
Oklahoma City, OK 73136-0415
(405) 425-7296
www.homelandsecurity.ok.gov/

Oregon
Oregon Military Department
P.O. Box 14350
Salem, OR 97309-5047
(503) 584-3991
www.oregon.gov/OMD/

Pennsylvania
Pennsylvania Emergency Management Agency/
Homeland Security Advisor
2605 Interstate Drive, Suite 380
Harrisburg, PA 17110
(717) 651-2715
www.homelandsecurity.state.pa.us

Rhode Island
State of Rhode Island
Emergency Management Agency
645 New London Avenue
Cranston, RI 02920
(401) 275-4333
www.riems.ri.gov/

South Carolina
South Carolina Law Enforcement Division (SLED)
4400 Broad River Run Road
Columbia, SC 29210
(803) 896-7001
www.sled.sc.gov/HSOfficeHome.
aspx?MenuID=HSOffice

South Dakota
Office of Homeland Security
118 West Capitol Avenue
Pierre, SD 57501
(605) 773-3478
http://dps.sd.gov/homeland_security/

Tennessee
Tennessee Department of Safety
and Homeland Security
312 Rosa L. Parks Avenue, 25th Floor TN Tower
Nashville, TN 37243
(615) 532-7825
Fax: (615) 253-5379
www.tn.gov/homelandsecurity

Texas
Texas Department of Public Safety
Mailing address:
P.O. Box 4087
Austin, TX 78773-0001

Street address:
5805 North Lamar Boulevard
Austin, TX 78752-4422
(512) 424-2000
www.tsdps.state.tx.us/index.htm

Utah
Department of Public Safety
4501 South 2700 West
P.O. Box 1411775
Salt Lake City, UT 84114-1775
(801) 538-3400
http://publicsafety.utah.gov/
emergencymanagement/

Vermont
Department of Public Safety
103 South Main Street
Waterbury, VT 05671-2101
(802) 241-5357
http://hsu.vermont.gov/

Virginia
Division of Law Enforcement
Standards, Policy and Homeland Security Manager
Department of Criminal Justice Services
1100 Bank Street
Richmond, VA 23219
(804) 225-3899
www.dcjs.virginia.gov/cple/ohs/

Washington
Washington State Patrol
Homeland Security Division
1519 Alaskan Way South
Seattle, WA 98134
(206) 389-2728
www.wsp.wa.gov/crime/homeland.htm

West Virginia
West Virginia Division of Homeland Security
and Emergency Management
Main Office and 24/7 Emergency Call Center
1900 Kanawha Boulevard East
Building 1, Room EB-80
Charleston, WV 25305
(304) 558-5380
www.dhsem.wv.gov/Pages/default.aspx

Wisconsin
Wisconsin Homeland Security Council
ATTN: WING-LGL
2400 Wright Street
P.O. Box 14587
Madison, WI 53708-0587
(608) 242-3075
http://homelandsecurity.wi.gov/

Wyoming
Wyoming Office of Homeland Security
Herschler Building, 1st Floor East
122 West 25th Street
Cheyenne, WY 82002-0001
(307) 777-4663
http://wyohomelandsecurity.state.wy.us/main.aspx

Guam
Guam Homeland Security
Office of Civil Defense
Ufisinan Difensia Sibet
221B Chalan Palasyo
Agana Heights, Guam 96910
(671) 475-9600
www.guamhs.org/

Puerto Rico
Homeland Security Advisor
P.O. Box 9066597
San Juan, PR 00906-6597
(787) 721-0435

Virgin Islands
USVI Territorial Emergency Management Agency
8221 Estate Nisky
St. Thomas, VI 00803
(340) 774-2244
www.citema.gov/index.html

American Samoa
Department of Homeland Security
American Samoa Government
Pago, AS 96799
(684) 633-2827

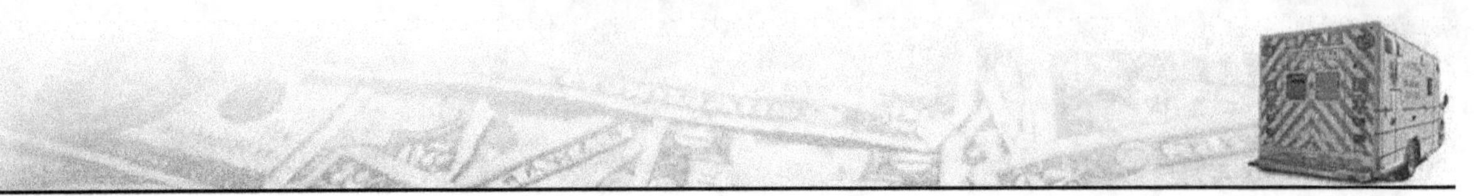

Appendix E: State Emergency Management Agencies

Alabama
Alabama Emergency Management Agency
5898 County Road 41
P.O. Drawer 2160
Clanton, AL 35046-2160
(205) 280-2200
Fax: (205) 280-2495
ema.alabama.gov/

Alaska
Alaska Division of Homeland Security
and Emergency Management
P.O. Box 5750
Fort Richardson, AK 99505-5750
(907) 428-7000
Fax: (907) 428-7009
www.ak-prepared.com

American Samoa
American Samoa Territorial Emergency
Management Coordination (TEMCO)
American Samoa Government
P.O. Box 1086
Pago Pago, AS 96799
(011)(684) 699-6415
Fax: (011)(684) 699-6414

Arizona
Arizona Division of Emergency Management
5636 East McDowell Road
Phoenix, AZ 85008-3495
(800) 411-2336 or (602) 244-0504
Fax: (602) 464-6356
www.dem.azdema.gov

Arkansas
Arkansas Department of Emergency Management
Building #9501
Camp Joseph T. Robinson
North Little Rock, AK 72199-9600
(501) 683-6700
Fax: (501) 683-7890
www.adem.arkansas.gov/

California
California Emergency Management Agency
ATTN: CSM Alex Cabassa
Training and Exercise Division
9800 Goethe Road, Box 46
Sacramento, CA 95827
(916) 324-9128
Fax: (916) 324-5929
www.oes.ca.gov/

Colorado
Colorado Division of Emergency Management
Department of Local Affairs
9195 East Mineral Avenue
Suite 200
Centennial, CO 80112
(720) 852-6600
Fax: (720) 852-6750
www.dola.state.co.us/ or www.coemergency.com

Connecticut
Connecticut Office of Emergency Management
Department of Emergency Management and
Homeland Security
25 Sigourney Street, 6th Floor
Hartford, CT 06106-5042
(860) 256-0800
Fax: (860) 256-0815
www.ct.gov/demhs/

Delaware
Delaware Emergency Management Agency
165 Brick Store Landing Road
Smyrna, DE 19977
(302) 659-3362
Fax: (302) 659-6855
www.dema.delaware.gov

District of Columbia
District of Columbia Emergency
Management Agency
2720 Martin Luther King, Jr. Avenue, Southeast
Second Floor
Washington, DC 20032
(202) 727-6161
Fax: (202) 673-2290
dcema.dc.gov

Florida
Florida Division of Emergency Management
2555 Shumard Oak Boulevard
Tallahassee, FL 32399-2100
(850) 413-9969
Fax: (850) 488-1016
floridadisaster.org

Georgia
Georgia Emergency Management Agency
935 East Confederate Avenue Southeast
P.O. Box 18055
Atlanta, GA 30316-0055
(404) 635-7000
Fax: (404) 635-7205
www.gema.state.ga.us

Guam
Guam Homeland Security/Office of Civil Defense
221B Chalan Palasyo
Agana Heights, Guam 96910
(671) 475-9600
Fax: (671) 477-3727
www.ghs.guam.gov

Hawaii
Hawaii State Civil Defense
3949 Diamond Head Road
Honolulu, HI 96816-4495
(808) 733-4300
Fax: (808) 733-4287
www.scd.hawaii.gov

Idaho
Idaho Bureau of Homeland Security
4040 Guard Street, Building 600
Boise, ID 83705-5004
(208) 422-3040
Fax: (208) 422-3044
www.bhs.idaho.gov/

Illinois
Illinois Emergency Management Agency
2200 South Dirksen Parkway
Springfield, IL 62703
(217) 782-2700 or (217) 782-2700
Fax: (217) 557-1978
www.state.il.us/iema

Indiana
Indiana State Emergency Management Agency
302 West Washington Street
Room E-208 A
Indianapolis, IN 46204-2767
(317) 232-3986
Fax: (317) 232-3895
www.ai.org/sema/index.html

Iowa
Iowa Homeland Security &
Emergency Management Division
7105 Northwest 70th Avenue, Camp Dodge
Building W-4
Johnston, IA 50131
(515) 725-3231
Fax: (515) 281-3260
www.iowahomelandsecurity.org

Kansas
Kansas Division of Emergency Management
2800 Southwest Topeka Boulevard
Topeka, KS 66611-1287
(785) 274-1409
Fax: (785) 274-1426
www.kansas.gov/kdem

Kentucky
Kentucky Emergency Management
EOC Building
100 Minuteman Parkway, Building 100
Frankfort, KY 40601-6168
(502) 607-1682 or (800) 255-2587
Fax: (502) 607-1614
www.kyem.ky.gov/

Louisiana
Louisiana Office of Emergency Preparedness
7667 Independence Boulevard
Baton Rouge, LA 70806
(225) 925-7500
Fax: (225) 925-7501
www.ohsep.louisiana.gov

Maine
Maine Emergency Management Agency
#72 State House Station
45 Commerce Drive, Suite #2
Augusta, ME 04333-0072
(207) 624-4400
Fax: (207) 287-3180
www.maine.gov/mema

Mariana Islands
CNMI Emergency Management Office
Office of the Governor
Commonwealth of the Northern Mariana Islands
P.O. Box 10007
Saipan, Mariana Islands 96950
(670) 322-9529
Fax: (670) 322-7743
www.cnmiemo.gov.mp

Marshall Islands
National Disaster Management Office
Office of the Chief Secretary
P.O. Box 15
Majuro, Republic of the Marshall Islands 96960-0015
(011)(692) 625-5181
Fax: (011)(692) 625-6896

Maryland
Maryland Emergency Management Agency
Camp Fretterd Military Reservation
5401 Rue Saint Lo Drive
Reistertown, MD 21136
(410) 517-3600
Toll-Free: (877) 636-2872
Fax: (410) 517-3610
www.mema.state.md.us/

Massachusetts
Massachusetts Emergency Management Agency
400 Worcester Road
Framingham, MA 01702-5399
(508) 820-2000
Fax: (508) 820-2030
www.state.ma.us/mema

Michigan
Homeland Security and
Emergency Management Division
Michigan Dept. of State Police
4000 Collins Road
Lansing, MI 48909-8136
(517) 333-5042
Fax: (517) 333-4987
www.michigan.gov/emd

Micronesia
National Disaster Control Officer
Federated States of Micronesia
P.O. Box PS-53
Kolonia, Pohnpei - Micronesia 96941
(011)(691) 320-8815
Fax: (001)(691) 320-2785

Minnesota
Minnesota Homeland Security
and Emergency Management Division
Minnesota Dept. of Public Safety
444 Cedar Street, Suite 223
St. Paul, MN 55101-6223
(651) 201-7400
Fax: (651) 296-0459
www.hsem.state.mn.us

Mississippi
Mississippi Emergency Management Agency
P.O. Box 5644
Pearl, MS 39288-5644
(601) 933-6362
Toll Free: (800) 442-6362
Fax: (601) 933-6800
www.msema.org

Missouri
Missouri Emergency Management Agency
2302 Militia Drive
P.O. Box 116
Jefferson City, MO 65102
(573) 526-9100
Fax: (573) 634-7966
sema.dps.mo.gov

Montana
JFHQ-MT
Montana Division of Disaster & Emergency Services
1956 Mt Majo Street
P.O. Box 4789
Fort Harrison, MT 59636-4789
(406) 841-3911
Fax: (406) 841-3965
www.dma.mt.gov/des/

Nebraska
Nebraska Emergency Management Agency
1300 Military Road
Lincoln, NE 68508-1090
(402) 471-7421
Fax: (402) 471-7433
www.nema.ne.gov

Nevada
Nevada Division of Emergency Management
2478 Fairview Drive
Carson City, NV 89701
(775) 687-0300
Fax: (775) 687-0330
www.dem.state.nv.us/

New Hampshire
Governor's Office of Emergency Management
State Office Park South
33 Hazen Drive
Concord, NH 03305
(603) 271-2231
Fax: (603) 271-3609
www.nh.gov/safety/divisions/bem

New Jersey
New Jersey Office of Emergency Management
Emergency Management Bureau
P.O. Box 7068
West Trenton, NJ 08628-0068
(609) 538-6050 Monday-Friday
(609) 584-5000
Fax: (609) 584-1528
www.ready.nj.gov

New Mexico
New Mexico Department of Homeland Security
and Emergency Management (DHSEM)
13 Bataan Boulevard
P.O. Box 27111
Santa Fe, NM 87502
(505) 476-9600
Emergency: (505) 476-9635
Fax: (505) 476-9695
www.nmdhsem.org/

New York
New York State Emergency Management Office
1220 Washington Avenue
Building 22, Suite 101
Albany, NY 12226-2251
(518) 292-2275
Fax: (518) 322-4978
www.semo.state.ny.us/

North Carolina
North Carolina Division of
Emergency Management
4713 Mail Service Center
Raleigh, NC 27699-4713
(919) 733-3867
Fax: (919) 733-5406
www.ncem.org/

North Dakota
North Dakota Department of Emergency Services
P.O. Box 5511
Bismarck, ND 58506-5511
(701) 328-8100
Fax: (701) 328-8181
www.nd.gov/des

Ohio
Ohio Emergency Management Agency
2855 West Dublin-Granville Road
Columbus, OH 43235-2206
(614) 889-7150
Fax: (614) 889-7183
www.ema.ohio.gov/ema.asp

Oklahoma
Oklahoma Department of Emergency Management
2401 Lincoln Boulevard, Suite C51
Oklahoma City, OK 73105
(405) 521-2481
Fax: (405) 521-4053
www.ok.gov/OEM/

Oregon
Oregon Emergency Management
Department of State Police
3225 State Street
Salem, OR 97309-5062
(503) 378-2911
Fax: (503) 373-7833
www.oregon.gov/OMD/OEM/index.shtml

Palau
Palau NEMO Coordinator
Office of the President
P.O. Box 100
Koror, Republic of Palau 96940
(011)(680) 488-2422
(011)(680) 488-3312

Pennsylvania
Pennsylvania Emergency Management Agency
2605 Interstate Drive
Harrisburg, PA 17110-9463
(717) 651-2001
Fax: (717) 651-2040
www.pema.state.pa.us/

Puerto Rico
Puerto Rico Emergency Management Agency
P.O. Box 966597
San Juan, Puerto Rico 00906-6597
(787) 724-0124
Fax: (787) 725-4244
www.gobierno.pr/AEMEAD/Inicio

Rhode Island
Rhode Island Emergency Management Agency
645 New London Avenue
Cranston, RI 02920-3003
(401) 946-9996
Fax: (401) 944-1891
www.riema.ri.gov

South Carolina
South Carolina Emergency Management Division
2779 Fish Hatchery Road
West Columbia, SC 29172
(803) 737-8500
Fax: (803) 737-8570
www.scemd.org/

South Dakota
South Dakota Division of Emergency Management
118 West Capitol
Pierre, SD 57501
(605) 773-3231
Fax: (605) 773-3580
www.oem.sd.gov

Tennessee
Tennessee Emergency Management Agency
3041 Sidco Drive
Nashville, TN 37204-1502
(615) 741-0001
Fax: (615) 242-9635
www.tnema.org

Texas
Texas Division of Emergency Management
5805 North Lamar
P.O. Box 4087
Austin, TX 78773-0220
(512) 424-2138
Fax: (512) 424-2444 or 7160
www.txdps.state.tx.us/dem/

Utah
Utah Division of Emergency Services and Homeland Security
1110 State Office Building
P.O. Box 141710
Salt Lake City, UT 84114-1710
(801) 538-3400
Fax: (801) 538-3770
www.des.utah.gov

Vermont
Vermont Emergency Management Agency
Department of Public Safety
Waterbury State Complex
103 South Main Street
Waterbury, VT 05671-2101
(802) 244-8721
(800) 347-0488
Fax: (802) 244-8655
www.dps.state.vt.us/vem/

Virgin Islands
Virgin Islands Fire Department
Road Town Fire Headquarters
P.O. Box 999
26 Fish Lock Road
Road Town, Tortola BVI VG1110
468 4267/4268
www.fire.gov.vg

Virginia
Virginia Department of Emergency Management
10501 Trade Court
Richmond, VA 23236-3713
(804) 897-6500
Fax: (804) 897-6556
www.vaemergency.com

Washington
State of Washington Emergency Management Division
Building 20, M/S: TA-20
Camp Murray, WA 98430-5122
(253) 512-7000
(800) 562-6108
Fax: (253) 512-7200
www.emd.wa.gov/

West Virginia
West Virginia Office of Emergency Services
Building 1, Room EB-80
1900 Kanawha Boulevard, East
Charleston, WV 25305-0360
(304) 558-5380
Fax: (304) 344-4538
www.wvdhsem.gov

Wisconsin
Wisconsin Emergency Management
2400 Wright Street
P.O. Box 7865
Madison, WI 53707-7865
(608) 242-3232
Fax: (608) 242-3247
emergencymanagement.wi.gov/

Wyoming
Wyoming Office of Homeland Security
Herschler Building, 1st Floor East
122 West 25th Street
Cheyenne, WY 82002
(307) 777-4663
Fax: (307) 635-6017
wyohomelandsecurity.state.wy.us

Appendix F: Success Stories

1. El Mirage (Arizona) Fire Department

El Mirage (Arizona) Fire Department purchased extrication equipment with $35,917 donated by a partnership of the Mahoney Company, an Arizona insurance company, and the Firemen's Fund Insurance Company. This equipment will enable El Mirage firefighters to perform extrications of people trapped in automobiles. Before this purchase, El Mirage Fire did not have an extrication capability and had to wait on firefighters to respond from Sunrise, Peoria, or Sun City to cut victims out of vehicles. The extrication equipment, which has cutting, spreading, and ramming attachments, was placed on a new fire engine which will be assigned to El Mirage's second fire station. This location was chosen because it is further from neighboring fire departments that frequently provide extrication services.

For information, contact El Mirage Fire Department, Howard Munding, Chief, (623) 876-4242.

2. West Virginia Pediatric-Restraint Systems

A chance meeting in a hallway led to purchase of pediatric-restraint systems for West Virginia ambulances, to ensure safe transport of children between 10 and 40 pounds. Vicki Hildreth, coordinator for Emergency Medical Services (EMS) for Children, had a short conversation with Christina Mullins, director of West Virginia Infant, Child, and Adolescent Health, which ended with Ms. Mullins asking Ms. Hildreth to submit a proposal for grant funding through the State's Injury Prevention Program. It resulted in an award of $230,000 through a Federal Maternal and Child Health Services Block Grant, which was enough to purchase a thousand child-restraint systems. The unit chosen for purchase rolls compactly, allows for efficient storage and retrieval, and is made of nontoxic, easy-to-clean vinyl for simple maintenance and reduced cross-contamination concerns. They have been distributed to the more than 200 West Virginia EMS providers and all practicing EMS personnel are being trained in their use.

For information, contact Vicki Hildreth at (304) 558-3956.

Website: www.wvstecs.org/operations/ems-for-children

3. The Bluff Creek (Louisiana) Fire Protection Territory

The Bluff Creek (Louisiana) Fire Protection Territory received a 2010 Georgia-Pacific Bucket Brigade grant of $10,000. It was one of 25 grants awarded from among more than 120 applicants in Georgia-Pacific communities in 23 States. Bluff Creek will use the money to replace personal protective equipment (PPE) and self-contained breathing apparatus (SCBA) for their volunteer firefighters. Fire Chief Allen McNabb said, "We truly appreciate the Georgia-Pacific grant. It will go a long way to help us replace some outdated safety equipment for our firefighters. This equipment will help us combat structure fires instead of just containing them in our rural community.

For further information, contact Fire Chief Allen McNabb at (225) 683-3827.

4. Savannah (Georgia) Fire and Emergency Services

Firehouse Subs Public Safety Foundation has awarded the Savannah (Georgia) Fire and Emergency Services nearly $20,000 to purchase water-rescue equipment. These resources will better equip firefighter/rescue technicians to operate at swift water and flood rescue incidents. Equipment being purchased includes wet suits, dry suits, masks, snorkels, gloves, boots, helmets, search lines, a variety of tools, dive-repair kits, and personal flotation devices.

For more information, contact Fire Chief Charles G. Middleton (912) 651-6758.

Website: www.savannahfire.org/cityweb/fireweb.nsf

5. Finger Lakes Migrant Health Care Project

The Finger Lakes Migrant Health Care Project received a Distance Learning and Telemedicine grant of $197,877 and leveraged $218,500 in matching contributions to establish the New York State Farmworker Telehealth and Learning Network. The grant funded telemedicine equipment for 13 rural health-care clinics in 10 of the most medically-underserved counties in the Finger Lakes Region. The project allowed residents' access to health professionals, specialists, and educational resources at five hub sites. Services provided include real-time telemedicine consultation with specialists, dental screenings, Web-based personal health records, and distance learning programs for residents and health-care professionals in rural areas of New York.

Website: www.rurdev.usda.gov/STELPRD4009127.html

6. North Charleston (South Carolina) Fire Department

The North Charleston (South Carolina) Fire Department will train all its firefighters as emergency medical technicians (EMTs), thanks to Assistance to Firefighters Grant (AFG) of $128,000 with a match of $32,000 from North Charleston. The department's medical responses have increased by more than 300 percent since 2004 and now make up more than two-thirds of total call volume. Some of the factors contributing to the increase include the growth of North Charleston in size and population, an increase in the number of seniors which places a higher demand on EMS, and the economic downturn that has caused many more people to be uninsured or underinsured. North Charleston firefighters will take 144 hours or training to become EMT-basic. "This grant will allow us to significantly improve our capability in providing initial medical response in coordination with Charleston County EMS," said Fire Chief Greg Bulanow.

For more information, contact Fire Chief Greg Bulanow (843) 740-2616.

Website: www.northcharleston.org/residents/Departments/Fire/default.aspx

7. Asher (Oklahoma) Fire and Rescue Department

The Asher (Oklahoma) Fire and Rescue Department received a grant of nearly $11,000 from the Thomas Gilcrease Foundation in 2009. It was a welcome donation and was used for equipment replacement and repairs. Jana Gilcrease, a Texas native, had recently moved to the area and was pleasantly surprised at the kind welcome she and her family had received from neighbors. She is Thomas Gilcrease's granddaughter and president of the foundation. "The foundation's main purpose is three things," she said. "We want to help out the Native American population, to help with education needs and to give in ways it will benefit the most people. The foundation awards scholarships at Oklahoma State University and gives to other charities and wants to help out in this area for its needs too." When she spoke to other foundation members, they were very receptive to her suggestion to provide the Asher Fire and Rescue Department with the contribution, which she and they believed would help a great number of people in Asher and the surrounding communities. "The fire department benefits so many people," she said. "There are always needs–not only fires but medical and others. We wanted to give the money so that it would benefit the most people and that's why we chose to give it to the fire department. The city and fire department need a little recognition for all they do."

8. Stockton (California) Fire Department

The Stockton (California) Fire Department won a Federal grant of $400,000 to buy a rescue truck, but that was not enough and the city, hit hard by budget woes, did not have $200,000 to complete the purchase. Food 4 Less covered the gap and the new rig is now in service with a Food 4 Less logo on the side.

Website: www.stocktongov.com/government/department/fire/default.html

9. Atlanta (Georgia) Fire Rescue

Atlanta (Georgia) Fire Rescue was awarded a $9.8 million Staffing for Adequate Fire and Emergency Response (SAFER) grant in 2011 to fund 75 new firefighter positions. "The grant accelerates our strategic plan for increasing staffing levels and providing better emergency medical system, rescue and fire suppression outcomes, while also enhancing firefighter safety and morale," said Atlanta Fire Chief Kelvin Cochran. Chief Cochran formerly headed the U.S. Fire Administration (USFA), which administers the AFG program. "That experience gave me a different approach to the competitive grant process," he said. The grant will solve several lingering staffing problems for the fire department, which has around 800 in field operations. The SAFER grant will allow Atlanta to staff each fire apparatus with four firefighters. Currently, over 66 percent of fire apparatus that respond to emergencies are staffed with only three firefighters. The new hires would all but guarantee that all trucks would have at least four firefighters. "That makes a significant difference in saving lives and minimizing property loss," Cochran said.

For more information, contact Fire Chief Kelvin Cochran (404) 645-7000.

Website: www.atlantaga.gov/Government/Fire.aspx

10. Alaska Code Blue Project

The Alaska Code Blue project is a partnership among the State of Alaska, the U.S. Department of Agriculture (USDA) Rural Development Division, tribal and municipal governments, community organizations, the Denali Commission, and the Rasmuson Foundation. The mission of the Code Blue Project is to provide essential EMS equipment by leveraging limited State dollars with funding from other nontraditional sources. Requests are reviewed and approved at the local and regional levels before being scrutinized by the State's Code Blue Steering Committee. Through the collaborative process, a single prioritized list of essential EMS equipment is established annually. During the first decade of the 21st century, $3 million in investments from Alaska was the catalyst for additional grants, yielding $15 million in capital projects.

With Code Blue funding, in 2009, the North Tongass Volunteer Fire Department (NTVFD) bought an ambulance and equipment for responding personnel, creating a viable advanced live support (ALS) response agency for its service area. This was NTVFD's second ambulance, allowing the department to provide EMS coverage from both its stations and have a backup, if one goes out of service. Jump kits for responding volunteers were also purchased.

In 2010, the Ketchikan Fire Department was awarded an Alaska Code Blue grant to purchase 2 automatic external defibrillators (AEDs) on the Port-of-Ketchikan and three AEDs in Ketchikan Police Department vehicles. An additional AED, with more advanced features, was placed in the Ketchikan Fire Department Incident Management vehicle. The cost of the six AEDs was about $10,000, with USDA picking up 55 percent, the State of Alaska contributing $2000, and the Ketchikan Fire Department the rest, about $2500.

11. Mecklenburg County, North Carolina

MEDIC, the emergency medical response organization for Mecklenburg County, is the busiest EMS agency in North Carolina, as it is located in the State's most populous county. A consortium of the two hospital systems in the county, Carolinas Healthcare and Presbyterian Healthcare/Novant, runs MEDIC through a dedicated contract with Mecklenburg County government. Although it operates under the same fiscal guidelines as other county-funded departments, MEDIC received only 32 percent of its funding from Mecklenburg County property taxes in 2010. The remaining 68 percent was generated through service fees paid by private insurance, Medicare/Medicaid, and self-paying customers. The county portion of costs per transport has dropped from over $400 in 1997 to less than $200 in 2010. MEDIC contracts with the Char-

lotte Fire Department and 18 other volunteer departments in the county for first-responder services, and the Charlotte Fire Department provides technical rescue services. In 2010, the focused cardiac arrest protocol, developed by MEDIC's medical director to create a highly-choreographed approach to working sudden cardiac arrest patients, yielded a successful resuscitation rate of better than 50 percent of cases.

12. City of Tracy, California

In 2010, the City of Tracy, CA considered implementation of an EMS service fee as a way to deal with budget deficits without having to cut service delivery. Like many California municipalities, Tracy has experienced significant declines in property values and consequently in property taxes collected. The proposal called for assessing $300 for EMS response but also included a $48 voluntary membership fee for Tracy residents, who then would not be assessed additional fees for EMS services they used. Tracy staff estimated that the fee structure would generate $455,050 in revenue for the city. However, the measure garnered opposition and Tracy City Council decided to delay implementation.

13. State of Virginia's Department of Health Rescue Squad Assistance Fund Grant Program

The State of Virginia's Department of Health (VDH) offers a grant program known as the Rescue Squad Assistance Fund Grant Program for nonprofit EMS agencies and organizations. Items eligible for funding include EMS equipment and vehicles, computers, EMS management programs, courses/classes and projects benefiting the recruitment and retention of EMS members. In 2011, the program awarded a total of $3,324,484 to 88 agencies. Most of the grants awarded require a local match, either 80/20 or 50/50 State/local. Applications must be submitted online via the VDH website (www.vdh.state.va.us/oems/grants/index.htm).

14. State of Florida's Department of Health/Bureau of Emergency Medical Services Matching Grants

The State of Florida's Department of Health/Bureau of Emergency Medical Services offers grants to counties and also matching grants to public and private urban and rural organizations that provide EMS. Applicants who are awarded matching grants are responsible for either 25 percent of all approved costs or 10 percent. Between 2007 and 2010, 153 grants were approved, for a total of $8,353,526.63 in State funds expended. Projects funded included communications equipment, computer equipment, AEDs, EMS training, EMS equipment, vehicles, and ambulance refurbishments. Eligible organizations must apply online at the Bureau's website: www.doh.state.fl.us/demo/ems/Grants/Grants.html

15. Merrimac (Massachusetts) Fire Department

The Merrimac (Massachusetts) Fire Department won Liberty Mutual Insurance's Be Fire Smart Pledge Program in 2010 with more votes than any fire department in the United States. For having the most people take the Fire Smart quiz, the Merrimac Fire Department won $10,000. Fire Chief Ralph Spencer said, "With the Be Fire Smart grant, we will further expand our award winning fire and life safety education and juvenile fire intervention programs. This grant will continue to allow the Merrimac Fire Department to provide the highest level of service to our customers, the residents and tax payers of the Town of Merrimac."

For more information, contact Fire Chief Ralph Spencer (978) 346-8211.

Website: www.merrimacfire.com/

16. East Providence (Rhode Island) Fire Department

The East Providence (Rhode Island) Fire Department received a $5.3 million grant in 2010 from the Assistance to Firefighters Station Construction Grant Program to update fire stations built in 1931 and 1954. Additional funding from an American Recovery and Reinvestment Act grant boosted the total available to

$6.4 million. Firefighters will move out of the stations to a temporary facility in an old school while construction is underway. The funding will pay for making the two stations complete two-story structures, renovate electrical, heating, and plumbing systems, pay for sprinkling systems in the two stations plus a third one, create additional space for administrative staff, increase the size of the apparatus rooms, and modernize the kitchens. Fire Chief Joe Klucznik said, "Naturally, we're elated…It means we're going to be working out of modern facilities."

For more information, contact Fire Chief Joe Klucznik (401) 435-7677.

Website: www.eastprovidenceri.net/content/666/738/746/770/default.aspx

17. Goodwin's Mills (Maine) Fire Department

Goodwin's Mills (Maine) Fire Chief H. Rodney Carpenter has been an effective grant writer for his small department. The community in York County broke ground for additions to its 7-year-old fire station in July 2010, paid for with $879,000 in Federal stimulus funds from a grant proposal Carpenter wrote, plus $100,000 in local money. The station needed to be upgraded; it lacked the sprinkler and ventilation and exhaust systems that are called for in Federal codes. The four-bay station can hold the department's trucks, but the addition will provide another bay to eliminate some of the double parking that's needed now. It will also provide office space, a kitchen, and sleeping quarters, even though the two full-time firefighters work only days. A third firefighter, approved by the town, will also work days, so there will be coverage 7 days a week. Chief Carpenter also landed a $128,000 grant for new breathing gear for firefighters.

For information, contact Fire Chief H. Rodney Carpenter (207) 499-2362.

Website: www.gmfd.org/

www.ingramcontent.com/pod-product-compliance
Lightning Source LLC
Chambersburg PA
CBHW081123170526
45165CB00008B/2526

* 9 7 8 1 4 8 2 7 7 1 7 7 0 *